A Woman's Guide to a
HEALTHY
STOMACH

A Woman's Guide to a
HEALTHY STOMACH

TAKING CONTROL OF YOUR DIGESTIVE HEALTH

JACQUELINE L. WOLF, M.D.

HARLEQUIN®

A WOMAN'S GUIDE TO A HEALTHY STOMACH

ISBN-13: 978-0-373-89223-5

The health advice presented in this book is intended only as an informative resource guide to help you make informed decisions; it is not meant to replace the advice of a physician or to serve as a guide to self-treatment. Always seek competent medical help for any health condition or if there is any question about the appropriateness of a procedure or health recommendation.

Library of Congress Cataloging-in-Publication Data
Wolf, Jacqueline L.
A woman's guide to a healthy stomach / Jacqueline L. Wolf.
   p. cm.
Includes bibliographical references.

ISBN 978-0-373-89223-5 (pbk.)
1. Digestive organs—Diseases—Popular works. 2. Women—Health and hygiene—Popular works. 3. Digestion—Popular works. I. Title.
RC806.W65 2010
616.3—dc22
2010035540

www.eHarlequin.com

Printed in U.S.A.

To the Loves of My Life:

My husband, David,

and

My daughters, Laura and Rebecca

CONTENTS

Introduction

*"Part of the secret of a success in life is to eat
what you like and let the food fight it out inside."*
—**Mark Twain**

This is a different kind of bathroom book. It's a book about bowel function—blowing the lid (so to speak) off the secrecy and shame surrounding female digestive ailments once and for all. This is a reassuring guide for women by a woman. It explains the causes and cures for our most embarrassing, urgent and common stomach problems. Wondering what those PMS cramps might mean? Always guzzling Pepto-Bismol before a big meeting? Read on.

Stomach ailments might just be the last great taboo in American culture. Women are the ones who suffer (I'm not just saying this—statistics back me up), and yet we're not whining about it! Seems silly, doesn't it? Bowel function is a necessary fact of life. We all go. But how many times have you hunched in the office bathroom stall, waiting for the boss to comb her hair and wash her hands, before letting loose with a massive explosion? Or wondered if your bad breath was caused by the onions you had for lunch—or something more sinister, like acid reflux? You're not going to cry about this over cocktails with your friends or coffee with your mother. No, it's easy to think everyone else is clean and pure, while you're the only woman alive with gas, acid, pain and cramping. But I'm here to tell you that your glamorous coworker with the designer clothes and perfect hair has stomach problems just like you, and it isn't always pretty.

It's during times like these, when things *aren't* pretty, that our stomachs become the center of our universe. The stomach is where we feel stress, nervousness, anxiety, pain. Just ask Freud. Yet, strangely, there aren't any books or websites that deal with stomach problems in a way that isn't completely satirical (ratemypoo.com) or incredibly technical (I won't bore you).

As a physician, I see this as a huge problem. Because when legitimate illnesses become shrouded in shame, they pose life-altering consequences for those who suffer from them. The repercussions range from the severe (undiagnosed ovarian cancer) to the annoying (planning out your driving route based on the nearest rest stop). By the time many patients reach my office, they've suffered alone for years, or they've been brushed off by doctors or told to take an over-the-counter medication.

Why? Bowel issues are hard to diagnose, thanks to symptoms that could really mean anything, and they're tough to talk about. They involve bad smells and strange noises. You might have constant gas, but who wants to go to a doctor complaining of humiliating farting? You might get constipated during your period, but would this move you to get a GI referral? No, probably not. That's where I come in. Consider this book your cheat sheet to bowel problems. This isn't a substitute for a doctor's visit—and please, if you have unusual symptoms, don't hesitate to get checked out—but this is a jumping-off point for women who need answers.

Just as important, I think it's helpful to recognize that men have it easier in this arena. (Sorry, guys.) As I've seen in my practice, stomach complaints are largely a "woman thing." Like it or not, men are more apt to boast about farting or joke about bathroom escapades. Prostate exams are a rite of passage that men fret about—and joke about, too. You can't turn on the TV without seeing a bronzed man in a hot tub singing the praises of Viagra. It's okay for men to talk about and make light of their issues! Why not women?

I'm not sure why. But I do know that when it comes to the stomach, women are more prone to suffer quietly, with physical and emotional consequences. We also suffer from issues, like PMS and endometriosis, that just don't affect men. And women are more likely than men to get gallbladder disease, autoimmune disease, irritable bowel syndrome (IBS) and constipation.

This shame and reluctance to seek help—or the tendency to seek it too late—have real-life repercussions. According to the National Institutes of Health, more than seventy million Americans suffer from digestive diseases. In 2004 more than 236,000 Americans died from digestive ailments. Over half of the deaths were due to cancer—colorectal cancer accounting for almost 40 percent of all cancer deaths. And in many of these cases, deaths could have been prevented if routine screening had been done and treatment had been sought at the outset of symptoms. In the United States, Canada and Northern Europe, women are more than twice as likely as men to seek the advice of physicians for changes in bowel function. In my gastroenterology practice at Beth Israel Deaconess Medical Center in Boston, 70 percent of my patients are women. And almost universally, these women feel alone and scared. There's no road map, no resource to reassure them that they're not imagining their problems or that they're going to be okay.

Instead, symptoms mean fear: Could my bloating mean cancer? Could my endometriosis mean that I can't get pregnant? I often find myself in the role of psychologist as much as gastroenterologist. And my message for the afflicted woman is this: you're not alone!

Each chapter in this book touches on the physical, emotional and social consequences of women's most common bowel conditions, from endometriosis to irritable bowel syndrome. In many cases, I highlight patients whose diagnoses are illuminating or particularly interesting (though for space's sake, they are abridged here, and out of concern for privacy, their names, occupations and other possible identifiers have been changed). These women wanted to tell their stories so that other women might know that, yes, we're all in this together. Indeed, while digestive dysfunction can point to serious problems, often it's a common ailment with a clear-cut solution. How reassuring for the millions of women scouring the Internet in secrecy, running to the bathroom between appointments and avoiding social situations for fear of an eruption to know that there's help. Each chapter also includes Q&As, designed to answer the most common questions I hear in my practice. You'll also find advice on what to ask your doctor and which medications are worthwhile (and which ones aren't), as well as nutrition tips.

So find a quiet corner—maybe your bathroom, even?—and start reading!

Chapter 1

How Uncouth: Stomach Shame

*"For marriage to be a success,
every woman and every man should have
her and his own bathroom. The end."*
—Catherine Zeta-Jones

Why are stomach ailments shrouded in shame and embarrassment for women? We all have to cope with them at some point in our lives—a volcanic explosion after a Mexican dinner, a knot in the stomach before a big job interview or after fighting with a spouse. This is normal, and speaks to how acutely stress and discomfort resonate in our stomachs.

But, for some of us, our stomachs are the center of our very being. Many of us live with constant constipation, diarrhea, indigestion, cramping… without relief and without answers. Often, we suffer quietly—scouring the Internet to self-diagnose or bouncing from doctor to doctor, trying to figure out what's going on. Even worse, many women simply figure that this is the way life has to be, and we don't get the help we need. It doesn't have to be this way!

Chances are, if you're reading this book, you have stomach issues or care about someone who does. This chapter offers basic information about the digestive system and the way it affects our lives when it goes haywire. I'll focus on an overview of common issues that I see in my practice every day and the typical effects these problems have on my patients' lifestyles. In the discussion below, when I use the word *stomach*, it's as a general term to describe any problem with the digestive tract below the chest (the abdomen or belly).

First things first: Bowel function is a fact of life, and it shouldn't be humiliating. Everyone goes to the bathroom! And when we go normally, we don't spend too much time thinking about it, right? However, when something goes wrong, it affects us deeply. For most of us, stomach function is a complete mystery, and we take it for granted when our digestive system works as it should. The flip side of this is that when it doesn't, we tend to panic.

Part of the reason for this panic is that it's socially unacceptable to talk about bowel problems. We have no problem moaning about a horrible headache or even PMS cramps, but if we're going to the bathroom constantly—or not at all—there's really no one to tell. This is ironic since children are often obsessed with excrement, both as a curiosity and as something that's a little naughty and taboo. However, as we become "socialized," we lose that obsession, or we just joke about it and brush it under the rug.

This isn't a new phenomenon. In the eighteenth and nineteenth centuries, the health of the bowels was often equated with the health of the soul. Bad odors indicated that rot lurked inside. Better to cover up the odors with perfume and sprays! This stigma hasn't evaporated with time. Bad breath, gas, belching—if we can produce such horrible smells and sounds, there must be something awry, possibly deep within our bodies. There's a reason people load up on mouthwash, chewing gum, perfume and floral-scented bathroom sprays!

Yes, our odors are best kept hidden. Society has deemed it unladylike to smell bad. It has not always been that way, of course. Here, it's instructive to examine the life cycle: Shortly after birth, a baby has its first bowel movement, beginning a lifetime "practice." The mother (or father), who usually changes the diapers, is aware of the baby's poops, gas and colic. She knows whether the baby's elimination is effortless or distressing, and there's no taboo in discussing this with the pediatrician. The toddler eventually becomes toilet trained. At this point, Mom and/or Dad have to be actively engaged in the elimination practices of their child, often spending hours in the bathroom, cajoling and pleading and teaching, plus many hours changing and washing sheets or clothes. We sit around with other parents and commiserate about our kids' bathroom woes. We have no problem sharing gory details about their bathroom habits—but would you really tell your fellow playgroup moms about your own constipation, gas or diarrhea? Not exactly something you gossip about over lunch. I think that kids must sense our discomfort with our own guts and exploit that unease. When children don't want to do something, they will often develop a gut ailment—abdominal pain, nausea or vomiting. It often takes an astute parent to recognize this action for what it is.

When then does "poop," gas or abdominal pain become a taboo subject of conversation? Why are they considered dirty, disgusting and embarrassing? As we get older, the bathroom becomes an intensely private place for women. In high school, guys shower and use the bathroom en masse in the locker room. Women, of course, require privacy and closed doors. As we get older and enter romantic relationships, a double standard begins to emerge: It's okay for your boyfriend or husband to joke or brag about his odors and noises, but it's just not something women do. Remember on *Sex and the City,* when Carrie Bradshaw hid from Big after accidentally farting at his apartment? Or when Charlotte York exploded with diarrhea on the girls' trip to Mexico and was mortified? Not pretty! These are lighthearted examples, but the core issue is troubling: Women do not like to discuss their stomach problems. And, as a result, they often suffer in silence for long periods of time.

In the United States, Canada and Northern Europe, women are more than twice as likely as men to ultimately seek the advice of physicians for changes in bowel function. Yet most women won't talk about their fears until pushed. They don't discuss them with their family members, significant others, friends or, in many cases, with their physicians; many women who come to see me skirt the reason for the visit until I pry it out, and once I do, I'm the first person they've told. Moreover, some stomach issues are woman-only problems, like endometriosis, ovarian cancer and that good ol' standby, PMS. Also, women just seem more likely to react sensitively to the issues that stomach ailments cause. The gut can become the center of a woman's life if she has belly pain, has too few or too many stools, has a hard time having a bowel movement, is worried about finding a toilet in time or has a problem with gas.

How can something so central to our well-being cause such embarrassment? The results of silence can be serious and life-altering: Some women I know won't leave their homes until they feel it's "safe" and they've done all they can do in the bathroom, lest they risk going to the bathroom in their underwear, or interrupting their commute or a work meeting to flee to the nearest toilet. Many wake up extra early and spend hours laboring in the bathroom, without anything to show for it except strained muscles. This may result in lateness to work, lost jobs, even depression and isolation. Many women are afraid to go out for dinner, to exercise or to travel. They fabricate elaborate excuses to stay home or just say they don't feel well. More than anything, they feel alone.

So *you* feel less alone, here are some of the most common complaints and concerns I hear from my patients. Can you relate? Short, easy-to-follow solutions are suggested for each concern. I'll address all these topics in detail in the book!

DR. WOLF'S TOP TEN MOST COMMON ISSUES:

1. "What if I can't make it to a toilet in time?"

Are you one of those women who can't go to the movies or the theater because you're constantly scrambling for a restroom? Or when you're on the highway, you're always searching for a sign pointing to the next exit with a possible bathroom. So many women end up altering their social lives or reducing them to nothing because it's just too embarrassing. Some of my patients have actually ranked public restrooms (note: hotels are usually safe bets) or highway rest stops because they so fear inopportune bowel explosions.

Try to use the bathroom before you leave the house. If you have frequent diarrhea associated with stress or IBS, you can take a half or a full dose of Imodium (loperamide) before you leave. (Check with your doctor if you have inflammatory bowel disease before taking loperamide.) If your bowel movement can be timed with your meals, eat earlier than usual, eat lighter than usual or delay eating until you get to your destination. If the diarrhea can be urgent and you can't hold it, you could wear an adult diaper (e.g., Depend). You're the only one who knows! It can give you security. Pack an extra set of underwear and clothing in your car, just in case.

2. "How can I go completely when my kids are banging at the door?"

I see countless mothers who suffer bowel issues yet can't fully evacuate, because their kids need attention. Parents rarely have time enough for themselves as it is. It's even tougher to be a mom when you're constantly running for the bathroom or spending extended periods crouching on the toilet.

Try to carve out some time for yourself. Often you need to relax before you can completely relieve yourself of feces. While on the toilet, it sometimes helps to put your feet on a stool or phone books or to bend down in order to give the poop a straighter, more direct way out. It is best to go sit on the toilet when you have an urge. It usually doesn't do any good to just sit down and wait.

3. "What if I fart while having sex?"

Not exactly the biggest aphrodisiac. Many women who suffer from gas, IBS and diarrhea often come to me with this fear.

Try to make this a nonissue. Avoid eating foods that cause gas several hours before you anticipate having sex. Go to the toilet before bed and see if you can eliminate gas or possibly stool. And *relax*. Don't occupy your mind with this worry, or you really won't enjoy the sex.

4. "How can I afford to buy all the supplements, vitamins and medications necessary to make my stomach feel normal again?"

Wander the aisles at any health food store, and you'll see plenty of pricey supplements promising stomach serenity. Unfortunately, you could spend a small fortune trying to compose your own stomach-soothing cocktail, with dubious results. How do you know what's worth it and what isn't?

Become an informed consumer. Television ads and company information are almost certainly biased toward the product, and it's often hard to assess the product's potential usefulness and whether the potential benefit for you justifies its expense. And the advice dispensed by a friendly salesperson might be uninformed. Read newspapers, magazines or online media for new studies. Manipulating your diet and lifestyle changes are cheap and could have major positive effects on your health. Lose weight if you are overweight, exercise, stop smoking and eliminate foods that could be causing symptoms, as discussed in the following chapters. Vitamin D and calcium, if not obtained in food, are important supplements for good health. Generic medications in most instances are cheaper and just as good as name-brand medications. Starting with a cheaper medication over a more expensive one is advisable. Probiotics have many health benefits, as will be described. Insurance will not pay for them and they can be expensive. All claims made about probiotics are not always based on studies. If you have IBS and want to take a probiotic, be sure to take a probiotic that includes a bifidobacillus. Other suggestions are given in the chapter on diarrhea.

5. "How can I afford to eat healthfully?"

These days everyone wants to be Mario Batali, trotting through bustling markets for the freshest produce. It's trendy, the food looks great, and you feel good about yourself after loading up your cart with shiny fruits and veggies, organic farmed fish and big bags of granola. But make a habit of shopping at fancy stores, and before long you've spent your entire shoe budget or mortgage payment. These delicacies might look pretty on the shelves—but buying them may not be the most money-wise thing to do. How can you eat well without breaking the bank?

Buy in bulk. Food is cheaper if you buy more than three pounds. Often joining a food co-op or a discount or wholesale market is cheaper. Specialty stores and organic food stores are more expensive. Think creatively. You can get much-needed protein from legumes and can add to them a small amount of meat and vegetables if you would like. Fish heads with the bones may be available from the fish market or store for a minimal cost. These can make a delicious soup. Many people don't think about eating a turkey unless it's Thanksgiving. However, turkeys are usually cheap and meaty and can stretch to feed many people. Buy items on sale and expand your palate. You might have to buy greens that you've never tried before—and you just might like them! You can freeze meat. Just make sure that you label it with the date and wrap it carefully. Check out the prices of frozen vegetables, as well as fresh vegetables. Make sure that you eat what is good for your condition.

6. "I'm always late for work because I can't get off the toilet!"

So many women I know end up running for the subway because they've spent more time sitting on the toilet than putting on their clothes. There's always that one last pitstop for the road—which sometimes turns into an hour of straining, grunting and pushing, to no avail. How many times can the "I was stuck in traffic" excuse hold up?

If the problem is that it takes you a certain amount of time after you wake up to finish your morning toiletry, wake up earlier. If the problem is that your cup of coffee sets you off, drink your coffee earlier, drink it on your way to work or wait until you get to work—whatever works. If the problem is that when you think of leaving home, you have to go to the bathroom, you should learn some biofeedback techniques. You may need to take an Imodium to get out. If you are taking fiber or a laxative during the day, try changing your intake schedule.

7. "I can't go out to eat—everything makes me sick!"

What fun is going to a restaurant when the menu is a land mine? Some acidic foods cause heartburn; other foods cause major gas. And forget about alcohol. How can you enjoy dining out if you have to order a prune salad or a milk of magnesia cocktail?

Try to scope out the menu ahead of time. The Internet is great for that. Know what you can eat. Call ahead and discuss your food preferences or needs with the chef. Often he/she will be very accommodating. If no accommodation will be made, go somewhere else. You should also carry a card noting what you cannot eat; this can be slipped to the waiter. If you have allergies or celiac disease, you absolutely need to make sure the chef understands your forbidden foods. Some restaurants will list gluten-free choices. If you don't think you can eat much at the restaurant, eat snacks at home before leaving and just order a few things (maybe just beverages) so that you can be social.

8. "My belly bloats up by the end of the day and I look seven months pregnant."

Do you have to open your zipper to feel comfortable or wear elastic-banded pants or find a loose-fitting dress to hide your figure?

The first thing to do is to try to eliminate any foods that cause increased gas, as this may cause bloating. Reducing stress may help. Learn biofeedback or meditation techniques to control stress and to reduce the physical accompaniments to stress. It is not known if strengthening the abdominal muscles will help prevent increased bloating, but it is worth a try and healthy, too. Probiotics may help. If bloating continues, try wearing loose clothing. Elastic-waisted clothes can expand during the day and be hidden by beautiful oversize sweaters. Body shapers can help control the belly bloat as well as the bulge. Most times no one else will notice. Don't broadcast your problem to your date or companion. Keep it to yourself.

9. "I can't control the noises my stomach makes!"

You're leading a board meeting, proudly giving a PowerPoint presentation, when suddenly your stomach erupts like Mount Vesuvius. Not the quickest route to a promotion. How do you rein it in?

This may require a course in biofeedback or a way that you find for controlling your emotions or stress. Sometimes there is nothing one can do. Everyone has had this problem at some time. Try to make light of it if you see someone staring at you. If your stomach makes loud noises when you're hungry, have a low-calorie snack before you go into that important meeting. Try to figure out what causes the noises, and then try to reduce the precipitating factor.

10. "I can't stop farting in public!"

You're sitting in class or at a meeting when suddenly a noxious odor seeps from your body. You avert your eyes or stare suspiciously at that annoying guy from accounting, trying to pin the blame on him. We've all been there at one time or another. But for some women, this kind of deviousness is a way of life.

First try to reduce the foods that cause gas. Do you need to eliminate milk products, gas-forming vegetables or other foods? If, after changing your diet, you continue to have this problem, you may need to be evaluated with breath tests. Could you have a parasite? Does simethicone (Gas-X) or enteric-coated peppermint capsules (do not use if you have heartburn) help? If you still have gas after changing your diet, consider a diet that eliminates fructose, lactose and other carbohydrates that cause gas, or a specific carbohydrate diet, or consider trying probiotics. If flatulence persists, you could try purchasing carbon filter underwear, which is available online. This could help absorb the odor, although not the noise.

The following questions and answers will help you understand your digestive system:

What is normal digestive function?

Remember these four facts:

1. **There's a wide range of normal!** Bowel movements can range anywhere from twice per day to three times per week. It's about understanding what's normal for *you*. If something feels off, don't ignore it. Bowel movements are affected by where you are in your menstrual cycle. Due to hormonal changes, many women have diarrhea or loose stools during their period, and constipation leading up to it. Some women actually look forward to their periods to empty out their bowels, not just to reassure them that they aren't pregnant.

2. **Stool form varies from person to person and from event to event.** No, it's not always a firm, one-inch-wide and six-inch-long form with a curled end. It can depend on your diet and fluid intake. "Paperless" was the ideal state of the bowels described by Heinrich Böll in *Group Portrait with Lady,* in which the stool comes out in one complete piece and leaves no residue behind. However, not many people have this kind of stool.

3. **Gas occurs in everyone!** In fact, we all pass between a half quart (500 mL) to a quart and a half (1500 mL) per day if we're normal. Some people make more gas than others. The amount depends on what you eat. It's normal to pass gas ten to twenty times per day.

4. **Check your meds!** Medications, including herbs and over-the-counter medications, can sometimes cause constipation, diarrhea or heartburn. But remember: most people have heartburn at one time or another—40 to 50 percent of Americans monthly and about 10 to 20 percent weekly.

How soon after I eat should food be expelled from the body?

You should eliminate that delicious sushi feast or burrito dinner within three to five days.

As a woman, how does my digestive system differ from a man's?

Well, we're all human, and for men and women, the digestive tract is made up of the same parts. It is a hollow tube that travels from the mouth to the anus. When food is ingested, it migrates from the mouth through the esophagus into the stomach and then into the small intestine. The small intestine comprises the duodenum (which is short), followed by the jejunum and finally the ileum. Then, remaining food goes to the large intestine or colon. The small intestine is twenty to thirty feet long, while the colon is three to five feet in length.

But there are differences. A woman's esophagus is shorter than a man's regardless of her height. Also, a woman's colon is often longer and "twistier" than a man's. This could contribute to more constipation, although no one has looked at the association of the number of bowel movements with the length of the colon. (It definitely makes it harder to perform a colonoscopy on women!) Finally, what's left of the food after its digestive tract travel is eliminated as waste.

Here's a handy diagram:

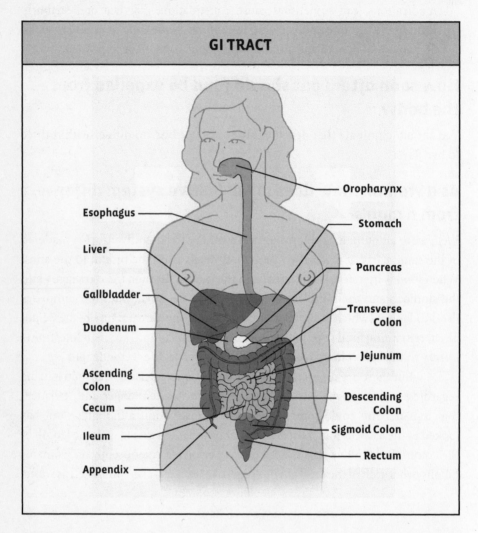

GI TRACT

Oropharynx

Esophagus

Stomach

Liver

Pancreas

Gallbladder

Transverse Colon

Duodenum

Jejunum

Ascending Colon

Descending Colon

Cecum

Sigmoid Colon

Ileum

Rectum

Appendix

Figure 1-1. The parts of the gastrointestinal (GI) tract with their locations are indicated. The food travels from the mouth through the esophagus, into the stomach and then out of the stomach into the small intestine. The small intestine is made up of (1) the duodenum, which is joined to the stomach in the upper right abdomen and then descends for a short distance before it heads across the belly to the left side, where it connects to the jejunum in the upper abdomen; (2) the jejunum; and (3) the ileum, which connects to the colon (large intestine) at the level of the cecum in the lower right side of the abdomen. There is no clear distinguishing characteristic marking the junction of the jejunum and the ileum. After the small intestine, the food travels into the colon at the level of the cecum. The main parts of the colon from cecum to rectum are indicated on the diagram. Bile is made in the liver, stored in the gallbladder and excreted into ducts. The common bile duct enters the small intestine in the duodenum. The pancreas makes digestive enzymes that are excreted into ducts, with the main duct entering the duodenum, usually with the common bile duct.

What does my digestive system do?

1. The saliva in the mouth moistens the food and starts the digestion process.

2. In the stomach, food gets mixed together, then broken down into smaller pieces, and digestion really starts in earnest with the help of acid and enzymes.

3. The small intestine breaks down the food substances with the help of more enzymes and proteins and absorbs nutrients and water into the body.

4. The large intestine (colon) takes out even more water to give you the stool form that we all know.

5. The intestines, particularly the colon, harbor helpful and harmful bacteria that produce nutrients and digest the food you eat.

6. The gut acts as a barrier to harmful substances and pathogens (organisms), keeping them out of the body.

7. The gut contributes to the immune response, making antibodies and fighting off disease.

8. The digestive system produces hormones and neurotransmitters that affect blood sugar, appetite and bowel function.

9. The digestive system eliminates indigestible substances as stool.

It's disgusting to think that I have "germs" inside me. What do these bacteria do?

There are one hundred trillion bacteria in your body. A good number of these are in your digestive tract. Most times they're working in perfect harmony with the body, helping to fight off bad bacteria and helping to digest your food. They make vitamins, as well as substances to help your body absorb the vitamins you ingest. But sometimes the ratio of good to bad bacteria becomes unbalanced, and symptoms or disease may become obvious. There is likely a healthy flora and it is unlikely to be the same in everyone. Lactobacilli and bifidobacteria are considered to be healthy bacteria.

What do bacteria have to do with weight gain and weight loss?

The more we learn about bacteria, the more we realize how much they influence our health. Take mice, for instance. Lean mice have different bacteria than obese mice, and it's not just a result of the type of food the mice are eating. When obese mouse bacteria take up residence in a lean mouse, the lean mouse gains weight, *even without eating any more food.* There are two main kinds of bacteria that seem to change with obesity in mice. The obese mice have fewer *Bacteroidetes* and more *Firmicutes.* Human bacteria can take up residence in the intestines of mice that are raised without bacteria. When one group of mice colonized with human bacteria was fed a high-fat, high-sugar diet, typical for a Western diet, the mice gained weight and grew more *Firmicutes* and fewer *Bacteroidetes.* In mice fed a low-fat, plant-based diet, the bacteria ratios were reversed. The change in bacteria to a high concentration of *Firmicutes* could occur in less than one day when the mice fed a plant-based diet were switched to a high-fat, high-sugar diet. It appears the bacteria may be important in people, too. In obese twins, there are fewer *Bacteroidetes* and more *Actinobacteria* than in lean twins. The bacteria that are present in obese individuals are more efficient at extracting calories from carbohydrates. Limited studies show that weight reduction in adolescents results in a change of bacterial flora. Furthermore, studies show that after gastric bypass surgery, patients had a change in their bacteria.

The types of bacteria associated with obesity may even be present before obesity occurs. Why is this important? The bacteria extract calories from carbohydrates and fats, and they stimulate the body to absorb these substances into the body and lay down fat. My patients have often told me that they can't lose weight, even though they are eating very little, or that they have even gained weight without overeating. These women were counting their calories, and they weren't snacking. Could it be that their bacteria are responsible? Could their intestines just be more efficient in absorbing nutrients (i.e., calories)? Can changing the gut flora result in weight loss? I'm speculating, but I believe the answer is yes. Only future studies will show if I'm right.

I hear a lot about probiotics and prebiotics. What are they?

Probiotics are living organisms thought to have good effects on one's health. They are similar to bacteria that are normally found in the intestines that do not cause disease and are often designated "good bacteria." The most common probiotics currently in practice contain a combination of *Lactobacillus, Bifidobacterium, Streptococcus* and other bacteria, or the yeast *Saccharomycetes boulardii*. Probiotics are present in some foods such as yogurt, fermented and unfermented milk, some juices and soy beverages. A large variety of probiotics are sold as supplements in capsule and powder form. Studies showing the utility of probiotics for immune health and gastrointestinal diseases are limited. Further discussion of the utility of probiotics for different conditions can be found in later chapters. However, it's important to keep in mind that no generalizations can be made regarding the effectiveness of an untested probiotic for a specific condition. Furthermore, it is unknown if a probiotic good for one person will be helpful for another.

Probiotics are often given to a person on antibiotics for protection against the development of *Clostridia difficile* bacterial infection. This infection occurs after the antibiotics kill off other gut bacteria that might keep the C. *difficile* in check.

Here's an interesting tidbit: One study in pregnant women published in preliminary form suggests that probiotics may be helpful in preventing pregnant women from developing obesity twelve months after the end of pregnancy. Further studies are needed to confirm this finding.

Prebiotics, meanwhile, are nutrients for the healthy bacteria. These nutrients, typically complex carbohydrates, are not digested by your gut and provide the bacteria a food source. Common prebiotics are inulin and oligofructose.

What are the most common stomach ailments?

Food poisoning is a biggie. It seems there's always a new scare out there: Don't eat raw eggs or undercooked chicken unless you want a nasty case of *Salmonella* or *Campylobacter*. Is a little taste of batter safe when you make a cake? (How many of us remember joyfully eating cake batter or cookie dough as kids?) There are warnings for raw meat, but fruit and veggies must be safe, right? Not so fast. Along comes the chance of food poisoning from contaminated spinach and tomatoes!

We'll all get food poisoning at one time or another. Other common ailments are heartburn (GERD), constipation and diarrhea, irritable bowel syndrome (IBS), and colorectal cancer. But don't panic. For most stomach issues, there's a logical and highly treatable explanation. In fact, about thirty-five million people have IBS, belly pain or discomfort with a combination of gas, bloating, diarrhea or constipation. I'll discuss this more in Chapter 3.

How do I know when to seek help?

Many people endure months or years of gas, constipation, diarrhea and/or abdominal pain. They live with it. But when is it important to seek medical attention to make sure that nothing is seriously wrong? No one wants cancer, of course, but colon cancer is usually curable if caught early. Therefore, even if you feel well and have no belly problems and no family history of colorectal cancer, it's important to do routine screening starting at age fifty (forty-five for African-Americans). And know the warning signs below that could indicate cancer or other GI problems.

WARNING SIGNS: CONSULT YOUR DOCTOR IF YOU HAVE ANY OF THESE SYMPTOMS

1. Rectal bleeding

Any rectal bleeding is abnormal and must be checked out. Although blood only on the toilet tissue when you wipe may just indicate a local cause, such as hemorrhoids, any blood should result in you seeking attention and being examined, usually via colonoscopy.

2. New onset of abdominal pain

There are many causes of abdominal pain, some of which require immediate attention. Usually, troublesome abdominal pain surfaces suddenly, with or without fever. However, it can be present for a long time under the radar before it becomes more frequent or severe.

3. Unintentional weight loss

4. Dehydration

Symptoms can include decreased urine production, thirst, dry mouth and eyes and dizziness.

5. New onset or worsening of diarrhea

Diarrhea springing from a virus or a bacterial infection often goes away on its own but sometimes requires further treatment if it is severe, with many watery stools (with or without abdominal pain), or if the symptoms persist for a long period of time. If the diarrhea continues for more than three days, consult your doctor.

6. Sudden change in the appearance of the stools or new and persistent constipation

Pencil-like or thin stools may indicate a narrowing in the colon.

7. The sensation that you have a mass, or a hard area, in the abdomen

8. Weakness and dizziness

This could indicate dehydration or anemia.

9. Prolonged fever

10. Repeated vomiting over a short period of time

11. Sudden onset of bloating that won't go away

Bloating can be due to gas, fluid in the abdomen, stool or sometimes a mass. If there is a blockage in the bowels, the abdomen can bloat up behind it, which usually causes pain. If you feel unusually bloated, consult your doctor.

Other problems, such as joint pain or a rash, can often be associated with gastrointestinal issues.

REMEMBER: You know your body better than anyone else. If you feel that something is wrong, check in with your doctor. Don't be afraid!

Chapter 2
Endometriosis and Feminine GI Troubles: Symptoms Every Woman Should Understand

"Man endures pain as an undeserved punishment;
woman accepts it as a natural heritage."
—Anonymous

W e women tend to feel stress in our guts more so than men, and we talk about it less, tending to bottle up our stress. We also have unique stomach issues, too, that simply don't affect guys. In this chapter I'll talk about women-only issues like endometriosis and infertility (5 percent of women suffer from endometriosis, and 25 percent of sufferers are infertile), and touch on some other female-centric ailments. You'll meet two of my patients, Marci and Susan, both of whom went through excruciating journeys to finally get proper treatment. Their cases are extreme but instructive: if one is properly armed, endometriosis is treatable, but it can be very hard to detect.

Women don't usually come to me thinking that endometriosis could be causing their gastrointestinal symptoms. In fact, many times women have either never heard of endometriosis or have had it in the past, and they certainly don't connect this problem with any gastrointestinal symptoms, thinking it's a purely gynecological issue. I'm not a gynecologist, and I don't treat women for endometriosis. However, I have seen many women with GI symptoms caused by endometriosis, often erroneously diagnosed as irritable bowel syndrome. Once I suspect the diagnosis, I can refer the woman to an expert who can decide what tests should be done and what treatment should be recommended for the problem.

When I tell women that they may have endometriosis, the women usually have a laundry list of questions. This is good. I find too many doctors who are not gynecologists know very little about endometriosis, and most gastro-enterologists don't even suspect it as a possible cause of symptoms. They often settle on a diagnosis before their patient has been properly heard. In order to understand what endometriosis is, it's helpful to review the anatomy of the gynecological tract and normal menstruation. (See Figure 2-1).

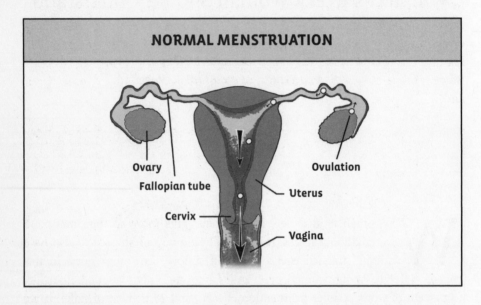

Figure 2-1. Menstruation is monthly bleeding from the uterus. The menstrual cycle is considered to start with the first day of bleeding. The next cycle begins at the time of first bleeding with the subsequent period. During the first half of the cycle (follicular phase) an egg in the ovary matures and the wall of the uterus thickens. At about day fourteen the egg is released from the ovary (ovulation) and travels through the fallopian tube to the uterus. The uterus lining continues to thicken (luteal phase). Then, if no fertilization of the egg with sperm takes place, the lining of the uterus is shed and discharged through the cervix and vagina as your period. Oral contraceptive pills will interrupt the menstrual cycle, but when menstruation occurs, it is normal.

Here are some common questions I hear from my patients:

Several of my friends have endometriosis, and one of them is having trouble getting pregnant. What is it?

Endometriosis occurs when the cells that line the wall of the uterus—which a woman should pass with each menstrual period—end up growing *outside* the uterus instead. During normal menstruation the uterine lining cells exit through the vagina (see Figure 2-1). But almost every woman also has retrograde menstruation in which some of the uterine-lining cells travel out the fallopian tubes and into the pelvic cavity, thereby tracing the egg's path in reverse from ovary to uterus (see Figure 2-2). This is called retrograde menstruation. When a person develops endometriosis these cells take up residence in the wrong places—sometimes even growing in and sticking to the bowel and other nearby organs—and can bleed with each period, causing pain and scarring. When this happens, a woman might get cyclical or constant abdominal pain. Usually, the endometriosis exists in the lower region of the pelvis, but it can creep onto other organs, too. (See Figure 2-3). Bowel endometriosis affects about one-third of women with endometriosis and can cause severe pain with bowel movements.

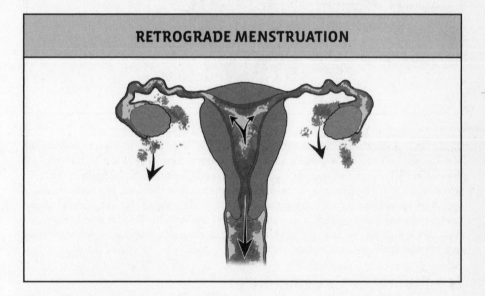

RETROGRADE MENSTRUATION

Figure 2-2. The menstrual cycle is the same as in normal menstruation. However, when the lining of the uterus is shed, it does not travel exclusively through the cervix and vagina. Some of the cells from the uterine lining travel upward, through the fallopian tubes and then out into the pelvis. This sets up the possibility for endometriosis to occur.

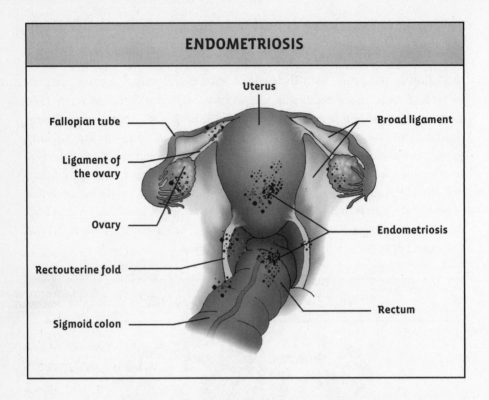

ENDOMETRIOSIS

Uterus

Fallopian tube

Broad ligament

Ligament of
the ovary

Ovary

Endometriosis

Rectouterine fold

Sigmoid colon

Rectum

Figure 2-3. Endometriosis occurs when the cells of the lining of the uterus take up residence outside of the uterus. The endometriosis lesions can be different sizes and vary from clear to red to black. This drawing shows the common areas where endometriosis occurs and the close proximity of the bowel to the uterus, which explains how endometriosis might end up on the bowel. The sigmoid colon is behind the uterus (toward the spine) and the rectum is behind the vagina. The rectouterine fold, also known as the uterosacral ligament, connecting the uterus and the sacrum, part of the spinal column, is a common place for endometriosis. The broad ligament connects the uterus to the wall of the pelvis and is in a perfect location for implantation of the endometriosis.

Is endometriosis common?

Approximately 5 percent of all menstruating women and girls suffer from endometriosis. However, for women who are infertile and can't get pregnant naturally, endometriosis can be the case 25 to 40 percent of the time! For teenage girls with very painful periods, endometriosis is the cause nearly 50 percent of the time.

If almost every woman has period cells that pass from the fallopian tubes, why don't all of us who still menstruate have endometriosis?

Women who do get endometriosis have several things going on: First of all, the cells have to stick to a place where they can grow and attract more cells. These cells then have to form blood vessels within the clump or implant—not an easy feat! Genetics and environmental factors also play a part. How your body reacts to these misplaced cells strongly contributes to the damage the endometriosis may cause.

Interestingly, tall, thin women are more likely to have endometriosis than short, heavy women. *Ahem!* This does not mean you should gain weight to decrease your risk for endometriosis. Obesity, of course, has many negative effects on your health. Some studies have shown that caffeine and alcohol intake also increase the risk of endometriosis, while smoking and exercise reduce the risk. (Of course, I'm not going to endorse smoking, either.)

I feel bloated and crampy all the time. My doctor says it's IBS. My coworker has endometriosis and told me I probably do, too, based on my symptoms. I'm starting to freak out. How do I know that I don't?

Endometriosis can be tough to diagnose because several of the symptoms—like diarrhea and cramping—can mimic irritable bowel syndrome (IBS). Don't settle for an IBS diagnosis, especially if you're having trouble conceiving. Here are some symptoms of endometriosis:

The Big Ds
- Dysmenorrhea—Otherwise known as debilitating menstrual cramps.
- Dyspareunia—Painful sex.
- Dyschezia—Painful bowel movements.
- Dysuria—Painful or uncomfortable peeing.
   PLUS:
- Recurrent miscarriage—Particularly pregnancies that end within two to three months.
- Nausea, vomiting and/or diarrhea—Particularly during PMS.
- Unusually long or short menstrual periods—a normal period should last between three and five days.

What is the best way to diagnose endometriosis?

The best way to make the diagnosis is by a surgical procedure called a laparoscopy. In this procedure, a small cut is made under the belly button, infused carbon dioxide distends the belly and a scope is inserted. The gynecologist or surgeon looks around to identify any raised blue, red or clear areas that could be endometriosis and examines the ovaries for cysts. Often these areas can be treated, and scar tissue can be cut (released).

Surgery is not always needed for the diagnosis. Other tests can be done but might not be as good for detecting very small lesions or for identifying scar tissue. These tests include an MRI (magnetic resonance imaging) scan of the pelvis, which can identify pelvic endometriosis; a CT (computerized tomography) scan with air or contrast in the rectum; and an air-contrast barium enema (to detect endometriosis on the bowel). An endoscopic ultrasound, in which an ultrasound device on the end of a scope is inserted into the rectum, can take images of the surrounding area and often see if there is endometriosis in the bowel wall. Traditional ultrasounds, with the ultrasound probe placed on the abdominal wall or in the vagina, and colonoscopies aren't particularly helpful for the diagnosis of endometriosis.

If endometriosis is so common, why is it so hard to pinpoint and diagnose?

The symptoms of endometriosis tend to mimic other issues. Women usually have different symptoms and may seek attention from a primary care physician, gastroenterologist or gynecologist. Abdominal or pelvic pain is a very common complaint. However, change in the bowels may also be a complaint. Endometriosis is not often the first thought for the internist or gastroenterologist.

Unfortunately, most doctors know very little about endometriosis. After all, isn't pain the norm with the menstrual period? *Suspicion* of the diagnosis is really what's needed to make a diagnosis. For more than forty-three hundred women reporting surgically diagnosed endometriosis who responded to a survey by the Endometriosis Association, the average time from the onset of symptoms until a woman sought medical attention was over *four and a half years,* and it took almost *five* years on average before the diagnosis was made. When symptoms started in adolescence, it took even longer before the correct diagnosis was made. And along the way, almost two-thirds of these women were told by one physician that nothing was wrong with them. Remarkably, 18 percent of women saw between five and nine physicians, and 5 percent saw *ten or more* physicians before endometriosis was diagnosed. Almost all of these women had pelvic pain and menstrual pain, and two-thirds experienced pain at ovulation.

Women often are totally accepting of having pain every month. But when is the pain *not* the norm?

Marci is a thirty-one-year-old lawyer who was referred to me for abdominal pain. For three months she had had an intermittent "pulling" pain deep in the pelvis, which seemed to come out of nowhere. This pain had grown more frequent, but she tried to ignore it—she had a frantic schedule, worked late all the time and just didn't have time to deal with it. She was also constipated, with very infrequent pellet-shaped stools. She chalked this up to not eating regularly, scarfing down lunch at her desk and sometimes skipping dinner or doing takeout. Her most recent menstrual period, she told me, was more painful than normal. Her pelvic pain also had begun to shift to the left side of her abdomen. She'd been sent by her gynecologist for a pelvic ultrasound and vaginal ultrasound, where a probe was inserted into her vagina. The probe in the vagina reproduced the pain. A small cyst found on her right ovary—a common finding in a young woman, often due to follicle formation—was the only finding.

Her days followed a specific routine, centered around pain: Every morning she would wake up hoping that *this* would somehow be the day that she'd miraculously be pain free. But her relief would be short-lived. After an hour, once she was up and about, the pain would start and get progressively worse as the day wore on. The pain radiated down her left thigh and occasionally shot down her left leg, though her back was pain free. She even dreaded lying down to sleep, because she knew she'd only lie awake in pain. Her only other complaint was tiredness. But then again, who wouldn't be tired given her lifestyle and lack of sleep? It was easy for her to make excuses.

Marci had suffered from colitis (an inflammation of the colon) five years ago, during law school, but a colonoscopy revealed that her colon had healed. She'd had a bout of appendicitis fifteen months ago, she told me, and her appendix was removed, which alleviated the sudden abdominal pain symptoms that she had at that time.

I reviewed her family history, and it was revealing: Her mother had endometriosis, and her mother's aunt had endometriosis. A positive family history for endometriosis, especially in her mother, made me suspect that Marci had inherited it as well. (In fact, heredity plays a part in 18 percent of cases.) Plus, Marci was an only child, which made me think her mother had had a hard time getting pregnant due to her condition.

When Marci came to me, she was very distressed about her previous physician interactions.

"My gynecologist told me that I must be under stress since my affect was 'flat,'" Marci said. "I told her that I was in quite a bit of pain and that it was definitely affecting my mood. She spent an hour asking me about the stress in my life, and then she told me I should go to a mind/body clinic! She said nothing was wrong with me." Marci's gynecologist brushed off her fears when Marci told her how hard it was for her to insert a tampon because of pain; her gynecologist also failed to biopsy her ovarian cyst. Clearly, by the time Marci reached my office, she was very frustrated—not to mention physically miserable.

My examination revealed only tenderness below the navel and on the left side of her belly. I examined her colon again by colonoscopy, and while I could see traces of her previous colitis, nothing could explain the pain that radiated down her leg. I also ordered an MRI for her spine, which was normal, and a pelvic MRI scan.

Once I got those results, I felt like we were getting somewhere: The pelvic MRI was abnormal, showing a small amount of free fluid in the pelvis and an area that suggested that there might be unexplained nodules. Her colon took some unusual sharp turns, which suggested that the colon could be stuck to adhesions (scar tissue). Adhesions are caused after surgery from inflammation (such as diverticulitis, "pouching" in the colon) or from the bleeding of the lesions of endometriosis. Taken together, my findings suggested the possibility of endometriosis.

The only way to be sure was to look in that area. This was done by a laparoscopy. (See above for a description of this procedure.) There it was— endometriosis with scar tissue, with involvement in her colon. Her lesions were cauterized (burned off), and she was started on Lupron (leuprolide) to reduce her estrogen levels.

Marci developed hot flashes due to a lack of estrogen. This was actually a good sign, as it meant there wasn't enough estrogen to stimulate the endometriosis to grow back. (Menopausal women experience this all the time.)

However, estrogen can be protective, especially against osteoporosis (loss of calcium in the bones, which makes them brittle). Marci's doctor thought that a low dose of estrogen could prevent osteoporosis without making her endometriosis rear its head yet again, as had been shown in studies. Therefore, her gynecologist prescribed an oral contraceptive pill. Unfortunately, within a week of starting the oral contraceptive, Marci began to suffer uterine bleeding. This was unexpected—Lupron should have stopped her periods entirely. (Marci's treatment isn't as strange as it might seem. Despite the fact that she did not have a period, her fertility wouldn't be permanently affected; once she stopped Lupron, her proper menstrual cycle and, hence, her fertility would return.) Even though this bleeding was clearly out of the ordinary, Marci assumed all was well, and she continued on her treatment. (Think this is weird? It isn't. I've had patients come to me after having ignored bleeding for years. Don't make this mistake!)

After three months of treatment, the bleeding became constant, so heavy that Marci went through a tampon every hour. For two weeks she went through a box of tampons every day. Her pain also became progressively severe. It was impossible to ignore any longer. On the advice of her gynecologist, she stopped the oral contraceptive pill. The bleeding stopped within twenty-four hours. The pain lessened after a few weeks but then returned, shooting down her left leg. Ibuprofen didn't help. The endometriosis was clearly back.

Now, five months after her first laparoscopy, an MRI scan showed a new lesion between her rectum and her spine on the left side. After she endured two months of exhaustion and increasing pain traveling down both legs, despite the Lupron, more laparoscopic surgery was considered.

She underwent another laparoscopy ten months after the first one, performed by a new gynecologist. Now she had endometriosis on the right ovary and additional endometriosis in her pelvis. Further evaluation showed new lesions in front of the spine and also between the rectum and bladder, which would certainly affect her bowels. She was told by the operating gynecologist that nothing else could be done for her—short of a hysterectomy with removal of the ovaries. At thirty-one, still hoping to start a family, she refused to accept this outcome.

At this point, it had been a year since she'd begun her journey. I had kept in touch with her via phone during her long saga. She now returned to my office for a GI follow-up. She told me her bowel movements were coming only once every three to four days, pelletlike and incomplete. Because constipation can accompany endometriosis, and especially given her new lesions, I thought the endometriosis was likely one reason why her stools were so odd. For the constipation, I recommended that she take Benefiber pills (guar) and add flaxseed (whole or ground), one to two tablespoons of each per day. I also started her on lactobacillus tablets. (A study in rhesus monkeys, which I talk about below, suggests that those with endometriosis had fewer lactobacilli. The benefits of lactobacillus supplements aren't definitively known for humans.)

Most troubling, though, was the pain that resulted from her deep endometriosis. She couldn't have intercourse or easily insert a tampon. I switched her to sulindac, a nonsteroidal drug, which I thought would work better than the ibuprofen.

Marci didn't like the sulindac, because it made her dizzy, and so she returned to ibuprofen for pain, sometimes taking up to eleven tablets every day. The ibuprofen made her more constipated, though it did somewhat address her pain. When she decreased the amount of ibuprofen at my strong suggestion, her constipation improved. She decreased the ibuprofen and put up with her pain.

Over the next eight months, Marci continued to have constant pain in her pelvis, although she was able to make it through her busy day, through a delicate dance of medication management. The pain caused exhaustion; she told me she'd crawl into bed at the end of a long day, praying for a pain-free night's sleep.

In spite of two surgeries, medical therapy and opinions from several surgeons, she continued to suffer. She had a choice: should she undergo hysterectomy, with the removal of the ovaries, which would likely take away all her symptoms? This is a difficult choice for a young woman, of course. Interestingly, if she did get pregnant in the future, the endometriosis would likely regress during pregnancy, though we don't know why. (Getting pregnant in the first place, however, might be difficult without IVF [in vitro fertilization].)

Before resorting to hysterectomy, she was seen by another gynecologist, who found tenderness on the pelvic exam consistent with endometriosis. He referred her to a gynecological-oncological surgeon, who was very experienced in complicated surgeries as he specialized in cancer in the female organs. The best time to operate was thought to be when Marci was bleeding during her period, since the endometriosis lesions could be more easily identified as they would be active. Therefore, she stopped further doses of the Lupron and surgery was scheduled when her estrogen levels were at their peak, meaning the endometriosis would be at its most visible.

The surgery was a success. The visible endometriosis lesions were cauterized (ablated) or removed, and the scar tissue was cut (lysed). Now on continuous birth control pills, Marci has been almost pain free for over one year.

How is endometriosis treated?

The goal of the treatment is to reduce pain, improve the chances for pregnancy and reduce any associated side effects from the endometriosis. Estrogen is a major factor in stimulating the endometriosis to grow. Therefore, treatment is aimed at interfering with estrogen stimulation.

Endometriosis can be treated medically and surgically. Your doctor should discuss the treatments with you in detail, as every treatment has possible side effects. Here's what he or she might suggest:

1. Oral contraceptive pills (cyclical or continuous).

2. Androgens: These are male hormones, like testosterone, which is the opposite of estrogen. A medication like Danazol increases testosterone and lowers estrogen. Beware—androgens can cause weight gain and masculinizing effects, like hair on your upper lip.

3. Gonadotropin-releasing hormones: These prevent the stimulation of the ovaries by your innate (natural) hormones and produce a low-estrogen environment. Lupron, leuprolide acetate, (given to Marci) is one such medication.

4. Progestins: Progestins stimulate progesterone receptors, helping to prevent ovulation and to lessen menstrual bleeding.

Are the symptoms the same for everyone?

The course of endometriosis varies from person to person. Marci's odyssey was severe. In fact, one-quarter of women don't have symptoms and might not even suspect a problem until they try to get pregnant and have trouble. Of those women with symptoms, pain can be mild or intense, like Marci's. The pain can be cyclical, occurring in relation to the menstrual cycle (often before your period), or it can be constant.

In up to one-third of women, the intestinal tract (usually the surface of the small and large intestines) houses these rogue cells. When endometriosis involves the colon or the small intestine, 40 percent of women experience constipation; 33 percent, diarrhea; and 5 percent, both diarrhea and constipation. Rectal pain during a bowel movement can be severe, and bleeding from the rectum can occur. It's little wonder that the diagnosis of irritable bowel syndrome is so often made in women who actually have endometriosis. A woman might even mistakenly be given the diagnosis of ulcerative colitis or Crohn's disease when she has bleeding from her rectum due to endometriosis.

Where does this pain come from?

The pain brought on by endometriosis is due to a variety of causes.

1. Irritation of the nerves occurs. When the endometriosis grows, it acquires a nerve supply and irritation of the nerves can cause pain.

2. Blood is an irritant. When the endometrial tissue bleeds due to hormonal stimulation, the blood can cause all sorts of inflammatory cells to migrate into the affected area.

3. Inflammatory cells release compounds that can cause pain. These substances stimulate nerves or cause pain directly by inducing inflammation, just like when you get a bad cut that develops inflammation around it. There are medications that act directly against the formation of these substances. These include aspirin, ibuprofen (Motrin), naproxen (Naprosyn) and sulindac (Clinoril). We use them to prevent menstrual cramps and for other aches and pains.

Do women with endometriosis tend to get other illnesses more often than women without endometriosis?

According to a survey conducted in 1997 by the Endometriosis Association, many conditions are found to occur more frequently in women with endometriosis. The most common coexisting conditions are allergies and asthma. Women with endometriosis also have a high prevalence of hypothyroidism, fibromyalgia, chronic fatigue syndrome, rheumatoid arthritis, systemic lupus erythematosus, Sjögren's syndrome (dry eyes and mouth) and multiple sclerosis.

Is there a diet that I can follow that can help the endometriosis?

Maybe. The data suggest that diets may help, but there are few good studies. Here's what I know.

In Japan a study showed that women who ate dietary soy isoflavones, which come from soybeans (tofu, for example), had fewer cases of advanced endometriosis, but the consumption of soy isoflavones did not affect the risk for early endometriosis. Soy isoflavones have estrogen-like activity but also can have weak antiestrogen properties. In this case, it was speculated that the antiestrogen properties decreased the risk for advanced endometriosis. Fish oil decreases the size of the endometriosis lesions in animals but hasn't been shown to decrease the risk for endometriosis or the size of endometriosis lesions in women.

Do probiotics help?

I wish I could tell you definitively, but I can't. The only evidence that suggests they might help is that the types of bacteria in the guts of monkeys with endometriosis are different from those in healthy monkeys. In monkeys with endometriosis, there are fewer lactobacilli. Whether or not endometriosis can be improved by taking probiotics containing lactobacilli just isn't known.

Are there any alternative treatments that work for endometriosis?

As any woman with severe endometriosis will tell you, it's worth trying anything! But there aren't any solid trials evaluating alternative treatments. In a self-report of 1,160 women responding to an Endometriosis Association survey, many different treatments were highlighted, including whole medical systems and energy medicine (including acupuncture, traditional Chinese medicine, candidiasis treatment, homeopathy and naturopathy, ayerveda reiki), mind-body medicine, biologically based therapies (including ingestion of dietary supplements, diet-based treatment and ingestion of herbs), and manipulative and body-based therapies (including exercise, chiropractic manipulation and massage therapy).

Biologically based therapies (use of substances found in nature) (52 percent) and manipulative and body-based therapies (based on manipulation and movement of one or more parts of the body) (41 percent) were commonly used. The self-reported improvements were 74 percent for therapy with mind-body medicine (techniques to enhance the mind's capacity to affect bodily function and symptoms) and 53–66 percent for therapy with many of the whole medical systems (complete systems of theory and practice), energy medicine (use of energy fields) and biologically based therapies. The manipulative and body-based therapies overall were reported to be less helpful, with 35 percent of women reporting improvement. However, without good studies, I can't recommend the alternative therapies just yet.

I have endometriosis. Do I have to worry about getting cancer, too?

If you have symptoms that are unusual or bothersome, certainly get them checked out. You should also have the routine recommended screening tests for cancers, such as PAP smears, mammograms and colonoscopies. There are some differences in cancer rates for women who have had endometriosis. In a very large study from Sweden that looked at the rate of diagnosis of cancer after a hospitalization for endometriosis, the overall rate of cancer was *not* increased. However, some tumors were slightly more common, and one—cervical cancer—was less common. The cancers that had about a 25–37 percent increase in incidence were ovarian, endocrine, thyroid, brain and kidney cancer, and malignant melanoma. Colorectal cancer was not examined but may be slightly more prevalent. Breast cancer was barely increased, possibly due to the fact that we screen so vigilantly these days.

Lately I've been needing to pee—constantly. Judging by my sex life, I'm sure I'm not pregnant. My mother wants me to get tested for ovarian cancer, just in case. I had no idea constant urination was even a symptom! Could I have it?

Probably not. Frequent urination is more often a symptom of other problems. It could be a symptom of a urinary tract infection or diabetes mellitus, or it might be associated with irritable bowel syndrome or interstitial cystitis. Burning with urination also occurs in almost half of all women with endometriosis. It's important to make sure that you don't have an infection by getting a culture of your urine and to make sure you don't have diabetes mellitus by having the sugar checked in your urine or blood. In the United States, females have a 1.4 percent lifetime chance of developing ovarian cancer. More than half of the deaths from ovarian cancer occur in women between the ages of fifty-five and seventy-four years. Still, it's important to be aware of the symptoms. If something feels unusual for your body, please tell your doctor! Many symptoms overlap with gastrointestinal issues. See your doctor if the following symptoms are constant or worsening:

- Bloating throughout the day, especially requiring a larger waist size on your pants

- Pelvic or abdominal pain
- Difficulty eating, feeling full quickly or weight loss
- Urinary symptoms (urgency or frequency)
- Frequent pain with intercourse

If I decide to get tested for ovarian cancer, what's going to happen?

It's important to remember that we want to rule out the zebras—or more unusual diagnoses—in the hopes of finding what we call horses, or more common ailments. Here's what you can expect.

Pelvic exam: Your doctor will feel your cervix, uterus and ovaries. She may do a Pap smear, which evaluates for cervical and uterine cancers or changes in their cells, but not for ovarian cancer.

Pelvic ultrasound: This will take a "picture" of your ovaries and analyze what might account for that full, bloated feeling. It does not involve any radiation, just sound waves. Usually part of the test involves putting an ultrasound probe in the vagina, which may show the ovaries better. If there are growths, the ultrasound can't always determine if these growths are likely to be cancer.

CT scan: This test uses X-rays to examine part of the body. It allows smaller problems to be detected. It visualizes the ovaries and uterus, as well as the bowel, lymph nodes and the spaces around them. A CT scan for ovarian cancer often includes an examination of the abdomen, as well as the pelvis. In that case, oral contrast is given to you to drink so that the bowel will stand out from the surrounding area. When the abdomen is examined, the liver, kidneys, spleen and pancreas are also seen. Often the radiologist doing the test will want to better visualize the blood vessels. This is done by an injection of dye into your arm. The dye contains iodine. So if you are allergic to iodine-containing substances, be sure and tell the doctor, as you will likely be allergic to the dye. Also, if you have any problems with kidney function, be sure to tell the doctor, as he or she might not want to do this part of the test.

MRI scan: This scan uses a magnetic field instead of X-rays to view the internal organs. It sees soft tissues very well. The best test is done with an enclosed scanner, where you'll hear a lot of banging. (If you're claustrophobic, speak up.) An injection of gadolinium (an element used as a contrast agent in MRI scans) is often done to see the blood vessels. Your kidney function should be confirmed as normal before you are given gadolinium, particularly if you have any problems that could affect the kidneys, like high blood pressure, diabetes mellitus, lupus, dehydration or kidney diseases.

Blood tests: A blood count (CBC) looking for anemia and liver function tests are commonly done. In fact, there is a blood test (CA-125) that had been touted to diagnose ovarian cancer. Unfortunately it is not a good screening test and has not been recommended as a routine screen in most people. CA-125 can be falsely high in someone who does not have ovarian cancer and falsely low in someone who does have ovarian cancer. On the other hand, CA-125 is often used to detect early recurrence of cancer in someone who had a high CA-125 with the original cancer and has had her ovarian cancer treated.

Laparoscopy: This is an even more precise test, in which a thin viewing tube (called a laparoscope) is placed through a small cut made in the abdomen. Using the scope as a guide, the surgeon takes a sample of fluid and tissue from the growth. These samples are then tested for cancer.

Every month around my period, I get bloated, I cramp and I have horrible diarrhea. I don't mean to be a big baby, but how can I deal with it without letting it ruin my life?

Well, first remember that you're not alone: about 85 percent of women suffer from some form of PMS each month, whether or not they have endometriosis. PMS, as defined by the American Congress of Obstetricians and Gynecologists, is "the cyclic occurrence of symptoms that are sufficiently severe to interfere with some aspects of life, and that appear with consistent and predictable relationship to the menses [menstrual period]." Only about 3 to 8 percent of women have severe symptoms. PMS symptoms may include upset stomach, bloating, constipation or diarrhea, appetite changes, mood disturbances, joint pain, headache and acne.

Changes in bowel habits during menstruation are reported by many women (34 percent in one study), and the symptoms are cyclical in almost 30 percent of women. At the time of menses, gastrointestinal complaints that women report are increased gas (14 percent), increased diarrhea (19 percent), and increased (11 percent) and decreased (16 percent) constipation.

One big tip—get enough calcium! A calcium supplement with vitamin D helps ease some symptoms of PMS. If you're between the ages of eighteen and fifty, you need at least 1000 mg of calcium per day. If you're older than fifty, you need 1200 mg. Eat plenty of fruits and vegetables, and get enough whole grains. Avoid alcohol and caffeine. They'll exacerbate your troubles.

For the bloating, try enteric-coated peppermint capsules before meals, although if they cause heartburn, you should stop. Loperamide (Imodium, one half to two tablets) could help treat or prevent the diarrhea, though too much can cause constipation. Pepto-Bismol is also worth a try—two tablespoons or tablets up to four times per day. Remember your stool and tongue may turn dark or even black after using it, but this is a harmless side effect.

I'm going to have a hysterectomy, and I'm worried that it might affect my bowels. I am already somewhat constipated. Will I get worse?

Probably not. Chronic constipation after a hysterectomy, unless it is an extensive operation for cancer, is not common. One study reported less frequent bowel movements, more laxative use, harder stools and constipation after hysterectomy, but this was not statistically more significant than in women who have not had hysterectomies. In more recent studies, no increase in constipation occurred in women without GI symptoms who underwent a hysterectomy. Furthermore no increase of IBS occurred after a hysterectomy in women without GI symptoms before surgery. Overall, movement of the stool through the colon does not change as a woman gets older, but the signal to let you know the stool is waiting to come out does decrease with age, unfortunately.

Can endometriosis come back?

Unfortunately, the likelihood that endometriosis will return is high. Five years after a patient has stopped medications to treat endometriosis, the recurrence rate is over 20 percent. Endometriosis and the pain associated with it can even recur after a successful ablation (cautery) or hysterectomy. The recurrence rate after surgery is higher when the ovaries (even one) are left or the endometriosis was severe, in which case it recurs in 30 to 47 percent of women. Of over eleven hundred women who had endometriosis diagnosed by surgery and who responded to a 1998 Endometriosis Association survey, 42 percent underwent surgical procedures for endometriosis at least three times.

You always have to be aware that recurrence is a potential problem. A cure of the endometriosis can only be assured if all estrogen, which can stimulate the endometriosis, has been removed. This occurs if a woman has both of her ovaries removed or goes through menopause. However, if one ovary is left behind after endometriosis surgery to prevent a woman from getting hot flashes or other symptoms related to menopause, the endometriosis can continue to be stimulated.

This problem often goes unrecognized. Women are told that they had a hysterectomy and that should "cure" the endometriosis. Not always! They don't know that a hysterectomy can involve removing only the uterus or removing the uterus with one or both ovaries. If you switch to a new physician, it's important for him or her to know right away what kind of hysterectomy was performed.

One such blunder happened with Susan, a forty-one-year-old woman who came to me with abdominal pain and diarrhea. Susan was a tough lady: a survivor of child abuse, she acted largely as a single mother to her handicapped son and her daughter while her husband traveled on business. She also worked full-time in a doctor's office. An avid athlete, she'd begun to curtail her activities because of her bowel problems, and her weight was fluctuating wildly. "I'm no use to anyone," she cried on our first meeting. "I feel like I'm in the twilight zone."

She was referred to me for a second opinion regarding her diagnosis of Crohn's disease. (Crohn's disease is a chronic inflammatory bowel disease. Crohn's can affect any part of the GI tract, and symptoms vary by patient depending on where the inflammation occurs. Symptoms can include constipation, diarrhea, abdominal pain, vomiting, weight loss or weight gain, and gastrointestinal bleeding.)

When I met Susan, she was clearly at a low: She'd have just a sip of water, then suffer diarrhea. She was having six to ten yellow, watery bowel movements every day, in spite of taking Imodium daily. Plus, she often woke up with fevers, which caused headaches and confusion. She was still able to work as an office manager at the doctor's office, but her weekends were consumed with sleep. Her eyes were inflamed, and her back was afflicted with arthritis. She also had hip, neck and leg pain, as well as sores in her mouth and a sore on her neck.

Susan had developed abdominal pain, rectal pressure and cysts in one of her ovaries after a hysterectomy for endometriosis a few years prior. She was treated with a large amount of Anaprox (a nonsteroidal anti-inflammatory drug like aspirin) for pain. She tried to tell doctors that the rectal pain reminded her of how she felt before her hysterectomy, but they brushed it off.

Shortly thereafter, Susan began to experience bloody diarrhea. She was admitted to her local hospital. A colonoscopy showed inflammation in her rectum at the end of her colon, and she was told that she had ulcerative colitis. She was treated with a steroid, prednisone, which has many possible side effects, such as diabetes, acne, and weight gain with short-term usage and bone loss and cataracts with long-term usage. Susan's symptom complex of abdominal pain, rectal pressure or pain and frequent stools occurred almost monthly. She was taking up to nine medications per day and still having a minimum of six bowel movements every day, too. She was also seeing an array of doctors, including a rheumatologist and an ophthalmologist, for side effects brought on by the steroids. Her weight was also fluctuating between 110 pounds (without steroids) and 150 pounds (with steroids). The steroid was eventually tapered down and stopped, and Susan was changed to a nonsteroid compound. She did well for a while, until she developed diarrhea, a whopping twenty-five times per day. She was treated with steroids yet again, despite the fact that she didn't even have colitis and had experienced side effects previously.

"I now have eye inflammation, arthritis and decreased calcium in my bones," she told me at our meeting. "Both specialists, a rheumatologist and an ophthalmologist, say it's from all the steroids and various other drugs. I am seeing a total of *six* doctors! Why can't I get better? The steroids are killing me. Still, I take them, then taper. My weight is going up and down. My old GI doctor is insisting that I have to comply and take the drugs. I tried to ask if it was endometriosis, but he's insisting it isn't."

Susan was somewhat lucky; thanks to her medical background, she knew what kinds of questions to ask. Still, she was seeing so many doctors and taking so many different medications that it was tough to get a clear picture of what was going on. And her doctor had overlooked the strong possibility of endometriosis, which I began to suspect.

I carefully considered her family history: a mother with breast cancer; two aunts with endometriosis, breast cancer, colon cancer and ovarian cancer; another with uterine cancer. Susan also had six older brothers, one of whom had Crohn's disease and another who had adult-onset diabetes mellitus and obesity. Susan herself had suffered from irritable bowel syndrome as a teenager.

Her examination was normal, save for a dark, raised round area on her neck and slight tenderness on the lower left side of her belly. Her blood tests showed borderline anemia. A repeat colonoscopy and upper endoscopy with samples of the bowel lining were completely normal. She was also tested for gluten allergy and did not have it. A review of the original rectal biopsy, when she was first hospitalized with bloody diarrhea, was consistent with an episode of infection.

These findings suggested that Crohn's disease was unlikely to be the cause of Susan's symptoms. Further history revealed that Susan always developed diarrhea whenever she was given the antibiotic clindamycin; not surprisingly, she had received clindamycin before dental work in the past, possibly just before the bloody diarrhea. Prior to her hysterectomy, she experienced rectal pressure, and around her menstrual period she would have diarrhea lasting up to eight days. The rectal pressure and diarrhea both resolved after surgery.

Because I really didn't think that she had Crohn's disease, I stopped all her medications. Off all meds, Susan reported that her diarrhea occurred up to six times per day after eating, three to four days out of the month, and was usually controlled by Imodium—a definite reduction from her previous bouts.

An abdominal CT scan was done one month after her second appointment, after all the records had been reviewed and the endoscopic procedures completed. This showed only a 2-by-2.5-centimeter round mass in the pelvis, close to the abdominal wall, the exact location where Susan felt her lower abdominal pain and where an ovarian cyst was found in the past.

Endometriosis as the cause for all her inflammatory symptoms was very unusual. Yet because Susan didn't appear to have Crohn's disease and was doing well off all medications, and because the CT scan showed an abnormality where Susan complained of pain, I referred her to a gynecologist who specialized in endometriosis. He performed laparoscopic surgery, which revealed a hemorrhagic ovarian cyst and scar tissue. The doctor didn't see endometriosis, though. He thought the pain was due to the ovarian cyst and the scar tissue. He drained the ovarian cyst and cut the adhesions. Her abdominal pain improved after surgery.

However, seven months post-surgery she began to develop severe left lower abdominal pain monthly with what she thought was ovulation. The pain was accompanied by sores in her mouth and fever. She continued to develop ovarian cysts, which would rupture. She had cramping abdominal pain at other times (different from the monthly ovulation pain), which she controlled with dicyclomine hydrochloride, an antispasmodic drug.

Her gynecologist started her on Loestrin, an oral contraceptive pill, to try to suppress ovulation. However, she could "feel" ovulation and did not think it had been suppressed, even though her diarrhea decreased. To treat the irritable bowel syndrome with which she had been diagnosed as a teen, she was started on the tricyclic antidepressant desipramine, which is helpful for decreasing abdominal pain and diarrhea.

But two years later she again developed severe lower left abdominal stabbing pain and fevers. Her blood pressure spiked to 146/96, and the Loestrin was stopped. (Hormones will increase your blood pressure at times.) A pelvic and abdominal exam showed tenderness on the left side, in the area of the ovary.

Susan's gynecologist performed yet another laparoscopy. Endometriosis was buried in the adhesions that were found. At that time, because of all her symptoms, the gynecologist removed both ovaries and tubes. Now, more than two and a half years after surgery, she has had no further abdominal pain or diarrhea. "I'm cured," Susan told me recently. "I have begun traveling again and exercising. I eat what I want when I want. I feel like a normal human being, and I'm a mom to my children." She thanked me for listening to her when no one else would.

If your ovaries are left at the time of surgery, endometriosis can come back. Women should discuss having their ovaries removed with their gynecologist, because doctors will often leave an ovary to prevent the menopausal symptoms that can occur after the ovaries are removed. If you have already had the children that you would like to have, removal of both ovaries may be a good option.

Why did Susan have to suffer for so long? Why was she misdiagnosed? The doctors didn't listen. Or, if they did, their preconceived notions did not allow them to ask the right questions and think outside the box. Perhaps they weren't knowledgeable about some of the side effects of medications, or perhaps they weren't knowledgeable about endometriosis. Either way, Susan was treated with the wrong medications—medications that caused a substantial number of side effects. Prednisone could have caused her osteopenia (bone loss), joint pain, weight gain, high blood pressure and possibly the skin infections. Her doctor almost put her on Remicade—a potent drug that suppresses an inflammatory substance called tumor necrosis factor.

Importantly, medications given to Susan by physicians played a major role in Susan's gastrointestinal illnesses. The Clindamycin, an antibiotic, likely caused the bloody diarrhea by inducing a *Clostridia difficile* infection, which was responsible for the bloody diarrhea. Nonsteroidal anti-inflammatory medications (NSAIDs), like aspirin, ibuprofen and naproxen, can also cause colitis. It would appear that one physician did not know what another physician had done. Susan tried to ask the right questions. She was more knowledgeable than most women, since she worked in the medical field. However, she wasn't calling the shots—her doctors were. And her questions were rebuffed.

For doctors, it's an easy trap to fall into. Once a diagnosis is given, it is often difficult to get that diagnosis changed. It's often easier to fit symptoms into that diagnosis if they seem to "mostly" fit rather than embark on an evaluation for a possible new diagnosis. When all the results don't fit—like in Susan's case—a new way of looking at the old and new problems has to happen. Susan tried to foster that way of thinking. However, the sicker she became and the more she was told she had to live with her problems, the more despondent and hopeless she grew. Some of her symptoms were not typical for endometriosis, and this led her doctors astray. Susan was misdiagnosed and on drugs she shouldn't have been on, and that caused side effects. I was at least able to correct the Crohn's misdiagnosis, take her off unnecessary drugs that could have caused substantial side effects and refer her to a doctor who could take care of her problem.

The moral of this story is be vigilant—you know your body best. If something feels wrong, say so. If you're left with more questions than answers after a doctor's visit, speak up. Get a second opinion.

WHAT YOU NEED TO KNOW ABOUT ENDOMETRIOSIS:

1. Endometriosis is a condition in which the lining of the uterus takes up residence outside of its proper location.
2. It is common in women.
3. It often mimics common gastrointestinal conditions, such as irritable bowel syndrome.
4. Many health-care providers are not adequately informed about endometriosis. Be your own advocate—ask whether you could have it.
5. It is difficult to diagnose with standard radiology tests and often requires an examination with a scope inside the pelvis or abdomen (laparoscopy).
6. There are both medical and surgical treatments for the condition, but recurrence is high if a woman still has her ovaries.
7. It is associated with an increased difficulty to conceive, but endometriosis seems to improve during pregnancy.

Chapter 3

"Do These Pants Come with an Elastic Waist?" The Truth about Gas, Bloating and Irritable Bowel Syndrome

"My philosophy on dating is just to fart right away."
—Jenny McCarthy

This chapter chronicles what happens when we can't fit into our pants, when gas escapes at inopportune times, when we have to beeline for the bathroom during an important meeting. We've all been there. But why does it happen? And, you're asking, why does it happen to *me?* After all, body odor is repellent, bizarre and unpleasant—especially for women. She might be gorgeous, smart and hilarious, but if she smells strange, well...all bets are off. Men, on the other hand, are sometimes allowed to smell rugged and musky. Guys work out and smell "ripe," and that's okay, maybe even alluring. Not so for women. So pity the poor woman who does suffer from regular flatulence. This is a mortifying situation, leading to low self-esteem and isolation, or at least complete humiliation.

In this chapter we'll meet Elizabeth, a thirty-seven-year-old art student who went to dozens of doctors in her quest to figure out why she was, in her words, a "gas factory." Her story is representative of those of many women I see—IBS can destroy a woman's life. By the time I met her, Elizabeth's sex life was lousy, her self-esteem was shot, and she'd been spending money running from specialist to specialist, who prescribed everything from antidepressants to antispasmodic drugs, when indeed she had irritable bowel syndrome. She was beginning to think she was crazy.

Elizabeth hardly seemed like a crazy, smelly woman: fragile and bird-like, weighing just one hundred pounds, she was pursuing a graduate degree in sculpture, which had been consistently derailed thanks to her ongoing stomach issues. By the time I met her, she had quit school and couldn't work. She told me that she had been "gassy" for as long as she could remember. She grew up in a traditional Asian home, where she suffered from frequent abdominal pain and the inability to control her gas. Her parents were mystified and ashamed—gassiness, in their opinion, was not an especially feminine trait. Her dad took to addressing her as "You, smelly girl!" and went so far as to tell her she mustn't be a girl, since she passed so much gas. "No man will ever want you like this," he told her.

Of course, this instilled a deep sense of unworthiness and translated into difficulty in intimate relationships. She spent her high school years isolating herself for fear of rejection. "I feel like my childhood and formative years were spent in the bathroom or in search of a bathroom," she told me when we first met. She also experienced a great deal of pain on a daily basis, which prevented her from connecting emotionally and participating in activities with her peers. Elizabeth's life, it seemed, had been defined by an ongoing waltz of pain and shame.

Elizabeth was seen by a physician, who brusquely told her to take a tranquilizer and see a psychiatrist. The psychiatrist helped her cope with some of her emotional baggage, but the sessions did nothing to relieve her symptoms. And what awful symptoms they were. She had severe, often debilitating pain and cramps in the abdomen and severe rectal spasms. These gave her the feeling that she needed to run to the bathroom to pass stool or gas, even if there was nothing to pass. She would end up in the bathroom all day, almost every day. This rendered her more or less housebound.

"Every day my main concern is, 'Uh-oh, do I have to run to the bathroom? Can I leave the house for ten minutes?' Wherever I go, I need to make sure I have easy access to a bathroom. At lectures I can hardly focus on what's going on. I'm plotting my escape route. Or else I show up late because I've been in the bathroom," she told me. At night she'd bolt awake with severe pain and rectal spasms, often spending hours on the toilet.

"I have trouble holding onto relationships because of this," she said wryly. "But I have great relationships with every bathroom in town." She did have a long-term boyfriend, but he was beginning to get fed up, too. It was hard for him to enjoy going anywhere with her when she was so clearly filled with dread about leaving the house. "I'm constantly preoccupied, and he's angry," she said. "My quality of life is in the gutter. My boyfriend is getting annoyed, and I'm not getting any sympathy from physicians. They think I'm exaggerating. It makes me not want to be around other people at all." She had begun to feel completely desexualized and had stopped having sex entirely. I felt immense empathy for this young woman whose life had clearly ground to a halt.

When I first met her, she was being treated with the antispasmodic medications belladonna and phenobarbital. She said the belladonna and phenobarbital helped a little, but only if she was not under stress. Stress made all her symptoms worse. Without the belladonna, she felt like she was a gas factory. She had to belch or "fart," or she would get a pressure in her stomach and lower belly. "I *know* how bad I must smell," she admitted to me.

Elizabeth was also coping with heartburn, despite the fact that she wisely avoided coffee, onions, mint and other irritants. Even a few bites of a totally bland food, like pudding, would make her feel full and give her heartburn. I suspected IBS and supplemented her medications with Pepto-Bismol, and for her heartburn she was given a prescription for pantoprazole, a proton pump inhibitor that stops acid. With some relief, she left my office.

Over the next month her upper abdominal pain and heartburn improved. However, in spite of the belladonna and phenobarbital, she had rectal pain that occurred throughout the day, lasting up to two hours at a time, which was debilitating. This was affecting her ability to take classes. A few minutes into a lecture she'd feel like she had to run to the bathroom. On the toilet, she might or might not have a bowel movement, but in either case the lower abdominal pain would not go away. I increased her phenobarbital with belladonna to four times per day. Because she still had severe rectal spasms, I tried adding an additional antispasmodic, hyoscyamine sulfate, which she could take as needed.

Over the next six months, her medications were modified. The belladonna and phenobarbital were changed to a long-acting hyoscyamine (Levsin), which would not make her sleepy or tired. Elizabeth continued to improve, with fewer episodes of pain. But when she was in pain, it was severe—she would remain in bed, unable to walk. Tricyclic antidepressants have been shown to help IBS pain, so I suggested she use a tricyclic antidepressant, Elavil (amitriptyline), in a low dose at night. (Studies have also shown that when anxiety or depression is treated, IBS will improve.) With the addition of Ativan (lorazapam) and Lamictal (lamotrigine) by her psychiatrist, her abdominal pain improved even more, although she did have some constipation as a side effect of these medications.

Despite some constipation, her quality of life continued to get better. The intensity of her cramping was much less, and she was able to do more. She still suffered from abdominal tenderness, but it didn't get any worse. I upped her Elavil even more, and her pain became more and more bearable. In the past she'd spend two hours on the toilet in the morning; this dwindled to just a half hour. Still, gas would return whenever she felt stressed, and she remained embarrassed about her odor, which she described to me as "silently fatal," particularly when she drank milk. Suspecting bacterial overgrowth (too many bacteria in her small intestines, which can result in abdominal pain and gas), I put her on a short course of tetracycline in addition to her other meds. (Tetracycline has been the traditional first choice of antibiotics for bacterial overgrowth, although rifaximin has fewer side effects and has been shown to help reduce gas and bloating. Unfortunately rifaximin is expensive and is not usually covered by most insurance plans until tetracycline fails.) To try to change the resident bacteria with new ones, I started her on a probiotic once the antibiotics ran their course. She also started drinking soy milk instead of cow milk. Elizabeth remained on the combination of medication that was helpful for her IBS, heartburn and anxiety/depression.

Today she says she feels a million times better. "After seeing Dr. Wolf, my symptoms didn't get better overnight, but at least emotionally, I felt that finally somebody was taking me seriously and not just writing me off as a hypochondriac. She started me on a set of medications to help me reduce my symptoms so that my daily life is not torture. My cramps used to be so bad, I'd break into a cold sweat. I couldn't talk. I'd be balled up in the corner of the cardio area in the gym for hours. Now that's not quality of life. But now it [the pain] is not that extreme. I feel a lot less embarrassed about being outside and being around people. I can function." In fact, she's now trying to finish her degree, and she hopes to get pregnant. "Instead of IBS managing me, I'm managing it," she told me at our latest meeting.

What is IBS?

Irritable bowel syndrome (IBS) occurs in about 14 to 25 percent of women, although most people with symptoms do not consult a physician and suffer in silence, due to shame or being told that they're overreacting. IBS also mimics many other issues, like endometriosis, so it's difficult to pinpoint right away. Elizabeth had a particularly bad and chronic case of IBS. Women are twice as likely to get IBS as men, and many of these women have suffered abuse, just like

Elizabeth had. Symptoms vary and are sometimes confused with those of inflammatory bowel disease (ulcerative colitis or Crohn's disease), which is a physical disruption or abnormality in the intestines that causes inflammation or damage to the bowel. With IBD, an X-ray, colonoscopy, endoscopy and so forth will show an abnormality; but with IBS they won't. IBS is a syndrome, not a disease.

The diagnosis of this syndrome is based on symptoms. There's no study that determines IBS; rather, the absence of abnormal X-rays and colonoscopy tests points to an IBS diagnosis. The diagnostic criteria have changed over the years but have always consisted of recurring abdominal pain and a change of bowel function—constipation, diarrhea or alternating constipation with diarrhea. Ask yourself this: Have you had recurrent abdominal pain or discomfort for three or more days per month? Has this been going on for at least the past three months? Does going to the bathroom help? Have the stools changed in frequency or form? And what about gas? Have you got it? If you have pain with the stool changes, you might have IBS.

Who gets IBS?

Those with a low quality of life are more likely to be afflicted because IBS may have a negative impact on one's overall well-being. Several psychological factors and childhood rearing practices have been reported to increase the risk of IBS. One study suggests that IBS in children is more common if the mother paid a lot of attention to complaints of illness and if children had many absences from school. A child with a parent or a twin with IBS is also more at risk for developing IBS due to genetic factors. Identical twins have a greater risk for developing IBS than nonidentical twins when one twin is diagnosed with the condition. The development of new onset IBS is associated with frequent visits to doctors, anxiety, sleep problems and somatic complaints (physical complaints, like aches, where the cause can't be found). Sometimes IBS crops up in the wake of a viral or bacterial infection; in this case, diarrhea is more likely to occur than constipation.

Some gut infections are more likely to cause IBS than others. After an infection with *Campylobacter jejuni,* a bacterial infection that comes from food poisoning from eggs and poultry, the risk of developing IBS is as much as 13 percent. It occurs less after *Salmonella* infections. But if you get antibiotics for your *Salmonella* infection, your risk almost doubles to over 17 percent, according to one study, which is why *Salmonella* usually isn't treated, except in the very young or elderly. Factors that increase the risk of developing IBS after an infection are being female, being younger than sixty, increased duration and (probably) intensity of the infection, psychological factors such as anxiety and depression, and smoking. Sometimes IBS is brought on by stress. Remember, your gut is hypersensitive and feels stress acutely. This is one way that it reacts.

What are the symptoms of IBS?

Without abdominal pain or discomfort, you don't have IBS! The pain or discomfort has to recur at least three days per month for three months in a row and also has to persist for six months. Secondly, the pain or discomfort has to be relieved by a bowel movement or be associated with a change in stool frequency (for you), or a change in stool appearance or form. This can be an increased or decreased frequency of stool or such changes in the stool as watery or mushy, hard, incompletely evacuated or requiring straining to evacuate. Bloating, gas and frequent urination, as well as the urgent need to urinate, are associated with IBS. Fibromyalgia and depression are also common in those with IBS. IBS is divided into subtypes, which are usually treated differently. These are constipation, diarrhea, mixed pattern (alternating between diarrhea and constipation) and undetermined. The constipation subtype is much more common in women, while the diarrhea subtype occurs equally in men and women.

Many women will tell you, though, that gas is their worst and most humiliating symptom. It's present in women with IBS 72 percent of the time.

My doctor asks me to describe my stool. I barely want to even look at it, let alone describe it!

Knowledge of the form your stool takes helps your doctor make an assessment about what might be happening in your intestines. A scale, dubbed the Bristol Stool Form Scale, has been developed just for this purpose. This scale helpfully describes seven major consistencies and shapes of stool, from little pieces (pebbles, rabbit droppings) to mush or liquid, so we doctors don't have to grope for vivid descriptions! Type 1 and 2 are hard to pass and occur with constipation. Type 3 and 4 are normal in the sense that they are more easily passed. Type 5 and 6 are soft and may come out with more force, and type 7 is diarrhea. Be sure to tell your doctor if your stool is bloody, that is, if blood is present either on the toilet paper or in the toilet, or is mixed in with the stool.

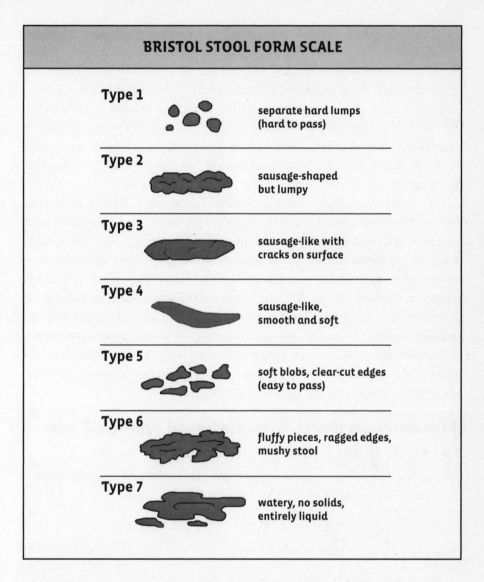

Figure 3-1. This chart shows the different types of stool form. Types 1 and 2 are hard to pass and are seen with constipation. Types 3 and 4 are considered normal (and ideal, by some people). Types 5 and 6 tend toward diarrhea. Type 7 is definite diarrhea with liquid and no form.

Can I get tested for IBS?

First things first: All women over age fifty and African-Americans women over age forty-five should have a colonoscopy to make sure there are no polyps or cancer. No matter what her age, any woman with unexplained rectal bleeding should have a colonoscopy to make sure there are no polyps, cancer or inflammation. Sorry, though, there are no accurate blood tests to make the diagnosis of irritable bowel syndrome. However, new blood tests have been developed to identify markers of substances (biomarkers) in the blood that are associated with irritable bowel syndrome. If there is a *high* suspicion of IBS and the test panel is positive—the tests have good predictability. Conversely, if there is a *low* suspicion of IBS and the test panel is negative, again, the test has good reliability. But if IBS is suspected and the test panel is negative, well, you still might have it.

Unfortunately, there are large numbers of people having these tests who could be overlooked and who do have IBS, or who are incorrectly told they have IBS. The blood tests sometimes used for diagnosing IBS include ones that are useful in making the diagnosis of celiac disease and inflammatory bowel disease; people with IBS could be told they have one of these other conditions. Other tests—such as stool cultures; routine blood tests looking for anemia or high white blood cell counts, which could indicate an infection or an inflammatory process; X-rays; or endoscopic tests—are useful in ruling out other conditions. However, for most people under fifty, they're not usually necessary. Celiac sprue (gluten allergy) (see Chapter 4) can be discounted by doing a blood test called tissue transglutaminase antibody.

If I'm diagnosed with irritable bowel syndrome how do I know that I got the right diagnosis?

You can't be 100 percent sure. Warning signs that would require other diagnoses to be eliminated with testing are:

- Onset of symptoms over age fifty
- Rectal bleeding
- New anemia
- Unexplained weight loss
- Worsening of IBS symptoms
- Poor appetite
- A family history of other diseases that affect the gastrointestinal tract
- Vitamin deficiencies

Reassuringly, studies have shown that the diagnosis of IBS does not usually change. After six months to six years after the diagnosis was made, only 2–5 percent of irritable bowel syndrome patients were diagnosed with another GI disease. With long-term follow-up of IBS, on average two years, about 2–18 percent of people developed worse IBS, 30–50 percent of patients had no change in IBS symptoms and 12–38 percent of people had a complete disappearance of their symptoms, for reasons unknown. If your symptoms change, tell your doctor.

Why do women bloat?

A great question that is still only partially understood! Most women do not have excess or an increased amount of gas, even though it might feel that way when you're squeezing into your favorite pair of jeans. The problem seems to be more that the gas doesn't move out of the small intestine like it should. And when it's there, watch out! Women with irritable bowel syndrome or other functional GI disorders feel it more. There is an increased sensitivity toward and awareness of what is happening inside the body.

However, not everyone who feels bloated *actually* has a larger belly. (In a study from Olmsted County, Minnesota, a quarter of the women studied reported they were bloated, but in actuality, only half of the bloated women had an increased circumference when measured!) Younger women are more likely to bloat than older women. Bloating is more common in the lower abdomen than the upper, unless someone has upper intestinal symptoms (like heartburn), whereby bloating can be anywhere in the abdomen or throughout the abdomen. With recurrent nausea, vomiting and pain (called dyspepsia), bloating is also more common in the upper abdomen. Men, meanwhile, are lucky. They bloat and distend only half as much as women. No wonder their pants fit better!

What about distension? Doesn't that mean you have more gas?

Again, not necessarily. Sometimes your ab muscles just aren't working right. The diaphragm comes down, the air and intestinal contents pool and the only way for the belly to go is *out*. Women can bloat if they get stressed. This may be due to a release of hormones or other substances that cause a change in the movement of the bowel or abdominal wall muscles, or just hypersensitivity. Another overlooked but likely cause of bloating is the bacteria in the bowel. The bacteria that are the normal inhabitants of the bowel are different in women with irritable bowel syndrome than those without it—one study shows decreased or absent species of lactobacilli in the intestines of patients with IBS. The fruits that are least likely to cause gas are white grapes, strawberries, blackberries, raspberries, pineapples and oranges. Fruits that are more likely to cause gas are prunes, pears, sweet cherries, peaches and apples. The bacteria in our guts break down the food products that get to the colon and can release rotten gas or compounds, like fatty acids, that could cause bloating.

What causes gas?

There's no getting around it: Gas is often a by-product of what we eat. We share our body with bacteria; in fact, there are ten to one hundred times more bacteria than cells in our body. These bacteria also inhabit our bowels. When the carbohydrates and fat within our food aren't broken down by the substances (enzymes) our bodies make and are not absorbed by the intestines into our bodies, the bacteria metabolize them and form gas as a by-product. For instance, if someone is missing the enzyme in the small intestine to break down lactose (the sugar in milk), she becomes lactose intolerant. The lactose travels down to the colon, and the bacteria break down the lactose and release hydrogen gas as a by-product. Fructose is poorly absorbed into the body unless it is accompanied by the sugar glucose. Therefore some fruits are more likely to cause gas than others. These fruits have less glucose than fructose. There are other poorly absorbed short-chain carbohydrates that will cause gas when broken down by bacteria. These are termed FODMAPs (Fermentable Oligo-, Di and Mono-saccharides and Polyols).

The oligosaccharides consist of fructans (wheat, onions, artichokes) that have long chains of fructose and galactans (legumes, cabbage, brussels sprouts). The disaccharides are lactose- (milk) containing foods. The mono-saccharides are fructose-rich foods (fruits). The polyols include sorbitol (added to low calorie foods). A more extensive list of the foods that may cause gas is found on pages 52–53.

Why do I feel stress—and, consequently, gas— so acutely in my stomach?

Strange as it seems, your gut has a brain. More than 95 percent of serotonin resides in the gut, and serotonin, coincidentally, helps modify mood. It is important for the movement of the gut and is a strong determinant of constipation, diarrhea and pain. But cortisol, the "fight or flight" hormone made in the adrenal gland, also causes increased movement in the gut and may be responsible for some of the increased gas when you're stressed. People with IBS also have a lower threshold for pain in the gut.

One of the tests done to look at how sensitive one's bowel is to noxious stimuli is to place a balloon in the rectum and distend it. If a balloon is distended in the rectum of an IBS patient, the tolerated volume of air in the balloon is less than that of a person without IBS. In IBS patients, when the balloon is distended in the rectum (ouch!) or if pain is anticipated, areas of the brain are activated. These can be seen on PET (positron-emission tomography) scans that examine blood flow. However, women's and men's brains light up in different places. In women, the limbic and paralimbic areas light up. These are areas that amplify pain. In men, areas that are more concerned with inhibiting pain (prefrontal cortex, insula, dorsal pons) light up. Men, too, experience intense pain; it just registers differently.

Why, oh why, does gas smell so disgusting?

Gas is actually a mixture of different gases. Only a minority cause odor (it's true!). The culprit for the rotten egg smell exuded by some gaseous odors is hydrogen sulfide. Other compounds like methyl mercaptan and short-chain fatty acids, like skatoles, can also impart a noxious odor to our released gas. Ironically, skatoles in small amounts have a flowery smell and are found in several flowers and essential oils, such as orange blossoms and jasmine (though I don't know of anyone who describes her gas as perfume-worthy). It's the sulfur-containing compounds that impart most of the bad odor to the gas. Depending on the type of bacteria residing in your gut and what food you eat, different gases and therefore different odors will be released.

What foods are notorious bloat causers?

Bananas, prunes, legumes, pretzels, cabbage, beans and raisins are big culprits. And, of course, dairy products produce bloating in lactose-intolerant people. Sorbitol, the sugar often found in chewing gum, causes gas, too.

How can I relieve gas and bloating symptoms while driving, or in a meeting, if I can't rush to a bathroom right away?

Whenever you're in for a long haul, visit the bathroom beforehand. And avoid those aforementioned foods that are likely to cause gas and/or bloating. If the trip or meeting is long, take bathroom and walking breaks. I'm also a big fan of relaxation techniques, such as meditation. Is that dragon-lady boss or tailgating driver really worth getting upset about, at the expense of your health?

Why are some women gassier than others?

Some women simply swallow air when they get nervous or eat. Some women aren't able to break down lactose. Still others have issues breaking down the sugar in fruits, called fructose. Some women also eat lots of beans and legumes, which have a nonabsorbable sugar known as raffinose, which is broken down by the resident gut bacteria that cause gas. Women like Elizabeth with irritable bowel syndrome often have gas and bloating—which is worse during the hormonal shifts around their periods.

What are some of the less common—and more severe—causes of distension?

Fluid in the belly, known as ascites, is one cause. This fluid can appear for a variety of reasons, from cancer to pancreatitis (inflammation of the pancreas) to cirrhosis (scarring of the liver). Abdominal masses and abdominal hernias are less common causes of bloating and distension. But bear in mind, all these issues are relatively uncommon and can often be detected on physical examination or with radiological tests.

Okay, so what foods should I avoid to reduce IBS and gas symptoms?

For IBS symptoms (abdominal pain, diarrhea), you should first try to eliminate "CAF" (caffeine, alcohol, fatty foods). If this doesn't work and you still have IBS symptoms or gas, try eliminating lactose for one week (milk, ice cream, cheese products). Decrease your meal size and eat smaller amounts more frequently. If this doesn't work, try decreasing fruits, followed by insoluble fiber. Red meats also seem to be poorly tolerated by some people with IBS symptoms. If all these things fail, sometimes a dietitian can help with an elimination diet.

Here are some commonly troublesome foods for causing gas:

- Beans and lentils (humans don't have the proper enzymes to break these tasty treats down. Sorry!)
- Brussels sprouts
- Carrots

- Celery
- Onions
- Apricots
- Bananas
- Prunes
- Raisins
- Pretzels
- Wheat germ

Beano, an over-the-counter preparation of the enzyme galactosidase, may help prevent the gas formed from the digestion of beans and peas. Women with bloating should avoid lactose and see if it makes a difference in bloating and gas. If it doesn't, you could go back to lactose. Sometimes it is not just the lactose but the milk proteins that cause belly discomfort.

Warning! Foods with lactose:

- Milk
- Cream
- Cheese
- Butter
- Yogurt
- Ice cream

What if I change my diet? I'll do anything! Give up ice cream! Throw away the wine! Will that help?

Many people report that diets for IBS and gas help. Unfortunately, not all diets help all people. One diet that helps one person may make another person worse. If allergy testing reveals food allergies, then elimination of food triggers may help symptoms. It is important for everyone to try to associate symptoms with a possible food that may have been eaten within the previous six hours (usually during the most recent meal). Keep a food diary! If there isn't a pattern, and allergies aren't found, there are certain things you can do to try to see if they help. Irritants to the gut, causing an increase in stimulation of gut movement and perhaps pain, are coffee, caffeine, alcohol, fatty and fried foods, and large meals in general.

As I mentioned before, some people get gas and pain from lactose, fructose or sorbitol. Insoluble fiber, mainly found in whole grains, raw vegetables and fruits, are often rough on the stomach. This may be due to the bacteria in the bowel eating up some of the fiber that gets to them or the stimulation of the bowel to have more contractions. However, it's hard to predict which foods will be tolerated and which won't.

I need help and I need it now. Can't I just run down to the local drugstore?

For gas, a simple treatment, such as simethicone (Gas Relief, Gas-X), works wonders. To reduce spasms, try enteric-coated peppermint capsules, sold under the name Pepogest at nutrition stores and specialized health-food grocery stores. Charcoal also binds the gas, and you can buy it as CharcoCaps. Keep in mind that charcoal turns the stool black, so don't be alarmed when that happens to you. Also, be sure not to take CharcoCaps with other medications, as it can bind up the medications so that they may not be absorbed effectively into your body. Pepto-Bismol may also be helpful to remove or decrease some of the foul odors.

Can't I just wear a "filter" in my underwear, kind of like the filter in my air conditioner?

Panty liners and underwear designed to absorb odors are available on the Internet, and I think they work pretty well. One study tested how well different products worked. Carbon fiber briefs extract almost all the foul-smelling gases, while pads made of fabric-covered charcoal absorb 55 to 77 percent of sulfide gasses. Cushions with carbonized cloth might also help. Sexy they're not, but then again, neither is a foul aroma.

How is IBS treated?

I've listed some of the common medications for IBS below. Good studies showing the effectiveness of the medications in therapy are often lacking. Therapy is mostly aimed at treating the major symptom(s). I discuss constipation treatment more in Chapter 5.

Fiber: A material contained in plants that can't be digested by the small intestine, it bulks stool, treats constipation and improves global IBS symptoms. (See Chapter 5 for a more thorough discussion.)
- Psyllium (ispaghula husk; Metamucil, Konsyl)

- Calcium polycarbophil (FiberCon)
- Other compounds used: wheat germ, guar, pectin, flaxseed, methylcellulose
- Laxatives: effective for constipation (see Chapter 5)

Antispasmodics: Antispasmodic drugs relax smooth muscle, decrease spasms and improve abdominal pain. These drugs are selected for therapy when a person is having significant spasms with abdominal pain or bowel movements urgently after eating. These drugs include:

- Hyoscyamine sulfate (Levsin)
- Dicyclomine hydrochloride (Bentyl)
- Clidinium bromide (in Librax combined with chlordiazepoxide [Librium])
- Peppermint oil, available as enteric-coated capsules

Side effects of the above antispasmodics (excluding peppermint oil) include dry mouth, constipation, urinary retention and visual problems.

Antidiarrheal agents: These substances used to decrease the number or frequency of stools can be used for IBS on a routine basis or as needed for the diarrhea.

- Loperamide (Imodium)—one or possibly two tablets every day can help decrease bowel movements
- Diphenoxylate/atropine (Lomotil)
- Pepto-Bismol

Side effects include constipation (with or without bloating) and cramps. Pepto-Bismol can cause a black tongue or stools and possibly could interfere with absorption of other medications.

Antibiotics: A short course of antibiotics often improves bloating and overall symptoms of IBS. Antibiotics act by changing the bacteria in the bowel.

- Metronidazole (Flagyl) can cause metallic taste in the mouth, abdominal pain, nausea, numbness or tingling
- Rifaximin (Xifaxan) acts locally and is minimally absorbed if at all
- Neomycin
- Tetracycline can cause sun sensitivity

Antidepressants: Antidepressants can provide relief from abdominal pain and overall IBS symptoms. Tricyclic antidepressants are very effective, but can cause constipation and are therefore used more frequently for diarrhea.

In my experience they may also be helpful for chronic nausea. They are used in small doses and given at bedtime.
- Amitriptyline (Elavil)
- Nortriptyline (Pamelor)
- Desipramine (Norpramin)
- Imipramine (Tofranil)

Side effects include fatigue, constipation, dry eyes, difficulty urinating and palpitations. They should not be used if you have glaucoma or problems with retention of urine.

Selective serotonin reuptake inhibitors (SSRIs) are helpful for overall IBS symptoms and abdominal pain. They usually are used in people with anxiety or depression.
- Paroxetine (Paxil)
- Fluoxetine (Prozac)
- Citalopram (Celexa)
- Sertraline hydrochloride (Zoloft)

Side effects include weight gain, constipation or diarrhea, nausea, dizziness and fatigue.

Serotonin receptor antagonists and agonists: These block the receptor and agonists and stimulate the receptor, respectively.
- Alosetron (Lotronex; antagonist): It is effective for relieving global (overall) IBS symptoms in women with IBS with diarrhea. Its use is restricted to women with IBS with diarrhea who have failed other treatments because of the possible serious side effects of constipation and vascular damage to the colon (colon ischemia).

Chloride channel activators: Chloride channel activators cause the secretion of saltwater into the intestinal lumen (cavity). They are effective in relieving overall symptoms in women with IBS with constipation. The stools become softer.
- Lubiprostone (Amitiza)

Side effects include abdominal pain and nausea.

What about alternative meds?

There have been only a few well-controlled trials involving patients with IBS. In one well-designed study, people with IBS were given individualized Chinese herbs, a standard set of Chinese herbs developed for IBS or a placebo. At the end of sixteen weeks, the two groups of people getting the herbs had improved bowel function and feelings of well-being. After another fourteen weeks off the herbs, only those who had received the individualized Chinese herbs still had benefit.

Ayurvedic (traditional Indian) herbs have been shown to be beneficial in decreasing diarrhea in those with IBS and compared favorably to a standard treatment with ispaghula (fiber), clidinium bromide (antispasmodic) and chlordiazepoxide (anti-anxiolytic). In people with constipation-predominant IBS, Padma Lax, a Tibetan herbal medicine, was significantly better than placebo for improving abdominal pain, stool frequency, distension and flatulence, and general well-being. (Padma Lax contains a laxative—cascara sagrada—as well as other herbs.) STW 5 and STW 5-II (both liquid herbal preparations containing bitter candytuft, chamomile flowers, peppermint leaves, caraway fruit, licorice root and lemon balm leaves) treatment resulted in significant improvement of overall symptoms. STW 5 acts as an antispasmodic.

Probiotics are also helpful for the treatment of IBS, but not all probiotics work. If you're looking for a probiotic to help your IBS symptoms, select one that contains a *Bifidobacterium*. This will be on the label.

I always see those Activia yogurt commercials. If it worked for Jamie Lee Curtis, it can work for me, right? What's so great about it?

Activia yogurt contains a probiotic (*Bifidobacterium animalis* subsp. *lactis* DN-173 010, or BL Regularis). Yogurt with the live bacteria has been shown to increase the speed (time) in which food passes through the colon by about 20 percent when eaten for two weeks. This may help constipation, although the data are very limited and need to be confirmed. Other effects on the immune system haven't been shown.

How can my psychiatrist help?

Many psychotherapeutic approaches are as effective in the treatment of IBS, either alone or in conjunction with other therapies, as the standard therapies discussed above. When stress is reduced, IBS can get better. Many different types of psychotherapeutic therapies have shown utility for treating IBS. These include hypnotherapy, cognitive behavioral therapy, dynamic psychotherapy and multicomponent therapy. Because psychiatrists and other therapists specialize in specific types of therapy, the type recommended by your therapist will depend on her skill with the treatment options. Relaxation therapy has not been shown to help in the studies that were done; I believe in it, but we need more information to show that it really works.

WHAT YOU NEED TO KNOW:

1. Everyone makes gas and harbors it in their bowels, but the amount varies from person to person.

2. Certain foods are notorious fuels for gas.

3. The bacteria in the body are the primary source for gas as they use the food we ingest for their fuel, with gas as a by-product.

4. To reduce gas, try changing your diet.

5. Irritable bowel syndrome—a condition in which a person suffers with belly pain or discomfort, with a change of stools and often gas, for a good period of time—occurs in as many as one in four women.

6. Don't despair; many good treatments exist for IBS.

Chapter 4

When You Know Every Bathroom in Town: Diarrhea

"There are only two reasons to sit in the
back row of an airplane: either you have diarrhea,
or you're anxious to meet people who do."
—Henry Kissinger

W e've all been there—bolting for the bathroom right before the big exam or crouched in the fetal position after gobbling up too much fried food. Sometimes you need to go, and you need to go *now*. It's an incredibly unpleasant, panicky feeling, and so many women cope with this kind of urgency all the time. In fact, in the United States it's likely that you'll get a diarrhea-related illness every one to one and a half years. In 2004 there were over 2,300,000 ambulatory care visits, 450,000 hospitalizations and 4,300 deaths due to gastrointestinal infections, totaling over $1.7 billion in direct and indirect costs. In 2010 Dr. Scharff from the Produce Safety Project, an initiative of the Pew Charitable Trusts, estimated that foodborne illnesses cost approximately $152 billion annually in health care and other losses in the United States. So next time you feel like the only woman in the world who needs easy access to a bathroom, take heart.

There are many possible causes for diarrhea—food poisoning, irritable bowel syndrome or inflammatory bowel disease, celiac sprue—and most of them are highly treatable. In this chapter I'll discuss ways to keep the diarrhea demon at bay.

We'll also meet Katie, a twenty-one-year-old college student with celiac disease, and Julie, a middle-aged mom with similar issues. Celiac disease, an autoimmune disease of the small intestine, is a major cause of diarrhea. It's caused by a reaction to gliadin, a portion of the gluten protein found in wheat and many other grains. This disease has become incredibly common as tests detecting it have become more and more refined, and the incidence is increasing. In the past, people with celiac disease were told they had anorexia, bulimia or IBS. These days celiac disease is a routine diagnosis and highly treatable. We'll talk more about it later in this chapter.

First things first: What is diarrhea?

True, we all think we know what diarrhea is—it's pretty unmistakable, after all!—but here's how doctors define it:

- An increased number of bowel movements per day (up to two is normal, but may not be normal for you).
- Increased stool bulk and amount.
- Stool that looks unusual (watery, loose, mushy). This is often accompanied by a strong sense of urgency. You have to go *now!*

I sometimes notice that my stool isn't brown. Sometimes it's yellow or green; other times it's very dark. Is this weird?

Stool comes in every color of the rainbow, and these colors are often clues as to what's going on in your body and reflect what you've been feasting upon lately.

Yellow or green: Sometimes stool turns these colors with an infection, though not always. Bile has a yellow-green color when it's in the liver. It contains a substance called bilirubin that comes from the protein hemoglobin, which carries the oxygen in your red blood cells. As bile travels through the intestines, the enzymes break down the greenish bilirubin, forming a compound that is brown. If the bile moves through too fast, then the stool may take on a green or yellow hue. This is harmless. Food dyes, Jell-O, Kool-Aid or sports drinks could give the stool a yellow or green color. Foods with plenty of iron, like spinach, can also give your stools a greenish tint.

Extremely dark: This could indicate internal bleeding, but there are benign causes, too. Often iron, iron-containing foods, Pepto-Bismol, black licorice or even blueberries can darken your stool. Ask yourself if you've ingested any of those things recently. However, black stools that look shiny and/or smell bad (different from the usual, that is) could very well mean you are bleeding internally. If your stool is shiny and foul smelling, call your doctor right away.

Bright red: If you're not having your period, then you shouldn't see blood in your toilet. And even with your period, blood should not come from the bowels. Beets; a large amount of cranberry juice, tomato juice or soup; or other pink drinks can turn the stool a shade of red easily mistaken for blood. If you haven't been eating the above, call your doctor immediately. This could be a sign of a serious problem, like cancer or polyps. It could also be due to a tear in the anus, hemorrhoids, or an inflammation caused by Crohn's disease or ulcerative colitis.

White or very light-colored: This is very unusual. It *can* happen if you take a lot of Kaopectate or recently took barium for a medical procedure. However, if it occurs without an explanation, then it may indicate that your body is having trouble getting bile out of the liver and into the intestines. This can happen with a blockage of the bile ducts by a stone or tumor, or when there is a problem with the pancreas. Call your doctor.

Pale, greasy and smells bad: This is usually a sign of malabsorption. In other words, there's too much fat in your stool, because your body couldn't process it. Causes include:

1. Diseases in which the small intestinal cells that absorb the fat or bile acids are damaged (celiac disease, Crohn's disease).

2. Conditions in which the enzymes for breaking down fats have decreased or are not present (chronic or sometimes acute inflammation or scarring of the pancreas).

3. Conditions in which the bile acids necessary to absorb fat, which are made in the liver and stored in the gallbladder, have decreased (liver disease, gallbladder disease, bacterial overgrowth in the bowel, post-surgery if large portions of the small intestine were removed).

Floaters: Floating stools are gassy stools—the gas makes them float to the water's surface. This could be due to what you're eating (check out the gas-inducing foods I listed in the previous chapter). However, it also happens with malabsorption. If in addition, you're experiencing weight loss or abdominal pain, call your doctor. If the floaters continue for one month, and even if you're not experiencing weight loss or pain, also contact your doctor.

Clear or white discharge: This is most likely mucus, which can be present with IBS or colitis; if you know you have one of these conditions, don't panic about the mucus.

I thought food poisoning was supposed to be a twenty-four-hour "blowout," but I was sick for five days with fever, diarrhea and cramps. My doctor insisted that I had food poisoning. How can that be?

Food poisoning happens when the bacteria and/or toxins or parasites in food react in our bodies to make us sick. The duration of the symptoms depends on whether the illness is caused *only* by the toxin itself, or by the bacteria or viruses that multiply in the intestines and then cause the disease. Some illnesses caused by tainted food can even last for months. (Fear not. This happens, but it's rare.) Here's a quick food poisoning guide. The organisms in this first table form toxins in food, so they act fast once the food is ingested.

TABLE 4-1: FOOD POISONING DUE TO BACTERIAL TOXINS

Bacteria/Virus	Incubation	Food Source	Symptoms	Duration
Staphylococcus aureus	1–7 hours	Ham, salami, mayonnaise or other protein	Abrupt vomiting often associated with diarrhea and cramping	6–24 hours
Clostridium perfringens	8–24 hours	Beef, poultry, pork or another meat	Upper abdominal cramping and diarrhea (with or without blood)	24–48 hours
Bacillus cereus	1–5 hours	Refried or reheated rice, vegetable sprouts	Abrupt onset of vomiting and, in 30% of people, accompanying diarrhea	12–24 hours
Bacillus cereus	6–20 hours	Refried or reheated rice, vegetable sprouts	Abdominal pain and diarrhea and, in 23% of people, vomiting	up to 36 hours

Some organisms take a few days to multiply inside your body before they cause disease. When they do surface, they can make you feel sick for several days or weeks. If you feel weak, have a high fever, can't take in any fluids, are getting dizzy or suffer from chest pain, then you should call your doctor immediately. If your symptoms of diarrhea last more than seven days or if you pass blood, also call your doctor.

The common causes for food poisoning due to bacteria and viruses that multiply in your intestines are listed in Table 4-2.

Is there any way to speed up my recovery from a GI virus or food poisoning?

If the trouble is caused by a preformed toxin from the organisms listed in Table 4-1 above, you're out of luck. It needs to run its course. Make sure that you try to take in plenty of fluids, especially fluids with electrolytes (like those in sports drinks), to stay hydrated. For mild dehydration, try juices and fruit punch. Two bouillon cubes dissolved in water are also an excellent source for replenishing salt. Oral rehydration solutions have a more balanced salt and potassium concentration; they should be available at your pharmacy or as packets in a hiking supply store.

It might be hard to get fluids down if you're nauseated, but do the best you can. Stay away from lactose-rich foods and drinks for a few days, which could make your diarrhea and gas worse, even if you're not normally lactose intolerant. Finally, avoid high-sugar or carbohydrate drinks as they could make your diarrhea worse.

For most types of food poisoning, though, studies have shown that Pepto-Bismol, Imodium and probiotics can help.

I want to take Imodium to stop my diarrhea, but a friend thinks it's a bad idea and says that I need to just let it all out. He says the medicine will keep the toxins in. Is he right?

Partially. Diarrhea is your body's natural way of ridding itself of toxins. However, in noninflammatory diarrheas, like those caused by viruses or bacterial toxins, it's perfectly fine to use Imodium, Pepto-Bismol or Lomotil. In traveler's diarrhea, Imodium is better than Pepto-Bismol at reducing the frequency of stools. However, with bacterial infections in which bacteria could possibly invade

TABLE 4-2: FOOD POISONING DUE TO BACTERIA/VIRUSES

Bacteria/ Virus	Incubation	Food Source
Salmonella	6–48 hours	Water, poultry products, eggs, unpasteurized milk, shellfish, spinach, jalapeño peppers, peanut butter, pistachios, packaged cookie dough
Campylobacter	1.5–10 days	Poultry products, eggs, unpasteurized milk and shellfish
Shigella	12–168 hours (av 72 hours)	Raw vegetables and salads
Enterohemorrhagic Escherichia coli	1–8 days	Contaminated meat, most commonly rare ground beef, but also poultry, lamb, pork, raw milk or unpasteurized apple cider
Cryptosporidium	5–28 days (av 7–10 days)	Swimming water contaminated with stool, tap water, raw produce
Vibrio parahaemolyticus	2.5–96 hours	Seafood
Yersinia	4–10 days	Raw milk and contaminated water
Listeria monocytogenes	6–240 hours (av 24 hours)	Raw cheese and luncheon meats
Norovirus	18–72 hrs (av 24–48 hours)	Infected water, shellfish, infected surfaces
Giardia lamblia	1–45 days	Contaminated surface water

Symptoms	Duration	Other Source
Watery diarrhea (may be bloody), usually accompanied by abdominal pain, nausea, fever, chills and headache	Usually 1–7 days, but could last 14 days	Pet turtles can be a source. Antibiotics in the elderly or very young. Increased risk due to acid-suppressing medications or if immunosuppressed. This is the number one cause of food-borne illness in the United States.
Fever, malaise and/or headache, followed by colicky abdominal pain and then watery and foul-smelling diarrhea that may become bloody; often with flu-like symptoms	7–14 days	25% of people relapse in a week or two. Antibiotics may reduce the relapse rate.
Fever, followed by diarrhea and abdominal pain. Initially watery, voluminous stools, then more frequent smaller stools, abdominal cramps and rectal pain. Gross blood in about 40%. Respiratory symptoms common. Joint pain infrequent.	Usually 4–7 days, but may be up to 14 days	Most common in institutions (nursing homes, homes for the disabled). Very infectious.
Diarrhea, often bloody. Associated with nausea, vomiting and fever.	Up to 11 days (av 3–7 days)	Can be very severe and cause kidney problems in some people. Spread can occur in household members.
Mild diarrhea to severe and even chronic diarrhea with cramps, malaise and poor appetite	Usually clears in 14 days, but may last longer	Protozoan
Abdominal cramps, diarrhea and usually vomiting, fever and headache	2–3 days	
Diarrhea with abdominal pain and fever. Less commonly loss of appetite, weight loss, nausea and fatigue.	14 days, but chronic infection lasting for months is not uncommon	May develop pseudoappendicitis or mimic Crohn's disease. Some types cause arthritis and other autoimmune symptoms.
Diarrhea, severe headache, fever, nausea, vomiting, joint pain	Less than 2 days	High risk of miscarriage in pregnant women. Meningitis not uncommon.
Vomiting/diarrhea alone or together and fever in ½	12–72 hours	Infection can be from airborne source. Highly contagious.
Foul-smelling often fatty diarrheal stools with cramps, bloating and bad gas. Also weight loss and nausea.	At least 2–4 months. ⅓–½ of people chronically infected	Protozoan. ⅔ of people are asymptomatic.

the body (things like *Shigella* or *Salmonella*), there's a chance that if you use Imodium or Lomotil, you could prolong the diarrhea and make the organisms spread into the bloodstream. If you have fever, blood in your stool or bad abdominal pain you may have a bacterial infection and you should not use Imodium or Lomotil without consulting your physician. If you're being treated with antibiotics for dysentery (severe diarrhea with mucus and blood), adding Imodium (loperamide) can decrease your number of stools. For dysentery, you absolutely must take antibiotics—not just an OTC treatment like Imodium. Pepto-Bismol can decrease the duration of the symptoms of viral gastroenteritis.

Every time I take antibiotics, I get diarrhea! The diarrhea usually stops when I stop the antibiotics. Last time I got sick, though, my diarrhea continued even after I stopped my pills. Why?

Some antibiotics have diarrhea or nausea as a side effect. Erythromycin and amoxicillin/ampicillin are especially common offenders. When you stop the antibiotics, the symptoms often go away. However, in some people, because the antibiotics change the bacterial flora, there could be some residual diarrhea, even after stopping the antibiotics. Don't worry; if this is the case, this will improve after your normal bacterial flora reestablishes itself. As we previously discussed, IBS may also occur after many infections, particularly from bacteria, and this may cause prolonged diarrhea (see Chapter 3).

However, *Clostridium difficile (C. difficile)*, a spore-forming bacterium, can overgrow in the bowel after antibiotic use and cause severe diarrhea, or even colitis with blood and serious illness. In some people, it's very difficult to eliminate the bacteria from their body permanently. Any antibiotic can cause this condition. One of the most common ones, ciprofloxacin, is even used to treat infectious diarrhea. Prolonged diarrhea following an intestinal infection treated with antibiotics (currently or in the past) may mean that you have yet another infection, and that you now need another antibiotic.

People over age sixty-five are especially vulnerable to *C. difficile* infection; so are people who take gastric acid-reducing drugs like omeprazole (Prilosec) or ranitidine (Zantac). If at all possible, you'll be taken off the antibiotics and likely put on metronidazole (Flagyl). But be forewarned: Flagyl is notorious for causing nausea and abdominal pain; it can also cause a metallic taste in your mouth. Why can't life always be easy? Still, it's effective in 90 percent of people with mild disease and 75 percent with severe disease. If Flagyl fails to work, your doctor might try vancomycin. (It's very expensive, though, which is why it isn't tried first.)

Probiotics like Florastor (*Saccharomyces boulardii*) are also helpful in preventing the occurrence of *C. difficile* infection in people on antibiotics. This still remains to be proved, although many studies appear to support the idea.

I vacationed for a week in Acapulco and spent most of the time reading guidebooks perched atop my toilet. Is there anything I can do to prevent diarrhea when I travel?

First, know where you're going. Low-risk areas for developing traveler's diarrhea are the United States, Northern Europe, Australia, New Zealand, Canada, Singapore and Japan. (It's still possible, of course, to contract diarrhea in these places.)

In Asia (with the exception of Singapore), Africa (outside of South Africa), South and Central America, and Mexico, the risk of developing diarrhea is high and you absolutely must use safe water. Here's what I tell my patients before they embark on a big trip:

1. Make sure you brush your teeth with bottled water.
2. Do not use ice in your drinks; freezing won't kill nasty bacteria and viruses!
3. Stick to canned or bottled juices if at all possible.
4. In some areas bottled water is not available and it might be necessary to boil or purify the water. Make sure you know this before you arrive!
5. Eat only fruits that can be peeled, such as bananas, oranges or mangos.
6. Avoid raw vegetables unless they have been peeled.
7. Under all circumstances, steer clear of condiments that have been left out—there's a high chance of contamination.

I can't stress this enough: Water is key! Unsafe water is the most common cause of infection. Bottled water is usually safe, but make sure the bottle was sealed—not just filled up from a tap.

How can I maximize my chances for having a diarrhea-free vacation?

Pepto-Bismol, antibiotics and probiotics can each help prevent traveler's diarrhea. Take two tablets or tablespoons of Pepto-Bismol four times per day for prevention. (Note: It might turn your stool or tongue black and make you constipated.) If you have ringing in your ears, cut down the dose by half. Treatment should be limited to six to eight weeks.

Antibiotics are also good for the prevention of traveler's diarrhea, but widespread use runs the risk of allowing resistant organisms to emerge. I try to steer patients away from taking preventative antibiotics, since it's possible to stimulate the formation of antibiotic-resistant bacteria. Plus, antibiotics can cause many side effects. Cipro (ciprofloxacin), levofloxacin, doxycycline and rifaximin (Xifaxan, which is very expensive and often not covered by insurance) are commonly used to prevent traveler's diarrhea. Probiotics may be of some help but are not as effective as antibiotics or Pepto-Bismol. More studies are needed to evaluate their utility.

Let's say I'm already on my gorgeous, exotic vacation when diarrhea strikes. What then?

You can try to decrease the frequency of bouts with Pepto-Bismol, taking two tablespoons or tablets every half hour for four hours, which can decrease the time and severity of viral gastroenteritis. If there is no blood, you can take a couple of Imodium. Plan ahead! If you're going to an area that is a hotbed of pathogens likely to give you diarrhea, see if your doctor will give you antibiotics that you can take if you get the diarrhea. This is especially important in elderly people or those with a condition that suppresses the immune system. Some areas in the world have bacteria that are resistant to certain types of antibiotics. Ciprofloxacin, levofloxacin and rifaximin are typically used for treatment.

Stay away from milk products initially, and if the diarrhea comes back after you reintroduce them, then withdraw them from your diet for one or more weeks. After infections, you may lose the substance in your small intestine (lactase) that breaks down the sugar (lactose) in milk. This may be temporary or permanent.

Most important, by treating symptomatic traveler's diarrhea when it happens, you can reduce the risk of contracting irritable bowel syndrome later on.

What can I do to stop the diarrhea and cramps that I always have before stressful events?

So many of my patients complain about stress-induced bowel problems! Studies show that acute (new) and chronic (long-standing) stress influences the intestinal barrier. (Interestingly, animals that are stressed are more likely to have bacteria escape from the bowel and cross the intestinal lining into the body and cause infection.) When you're stressed, the brain stimulates the release of epinephrine and cortisol, which causes the "fight-or-flight response." Your heart speeds up, your muscles feel ready to react and glucose increases in your blood. Stress also decreases the emptying capacity of the stomach; that's where that infamous "pit of my stomach" feeling comes from—your stomach can't completely eliminate its contents. Also, stress can increase movement in the colon, which causes the immediate urge to relieve yourself and produces abdominal pain.

Corticotropin-releasing hormone (CRH) is a major mediator of the stress response in the brain-gut axis. By blocking the release of CRH in the gut locally by the intravenous administration of a CRH receptor antagonist (not available for use now), there is improvement in gastrointestinal motility, visceral perception and negative mood in response to gut stimulation (stress) in IBS patients.

Finally, the best way to treat stress-induced diarrhea is to prevent it in the first place. Because as we all know, once the rumbling begins, there's really no turning back. Aside from trying to keep your life as stress free as possible, you can take some practical steps as well. Take some slow, deep breaths and try to get control over your pounding heart and fast breathing. If you're going to be in a situation in which you have limited bathroom access, try to eat lightly before going out. If you always get diarrhea before certain events (a big exam, speaking in public), try taking one Imodium before you go out to prevent the diarrhea from even getting started. Performers sometimes take a beta-blocker such as Inderal before walking onstage to keep their pulse in check, but I don't recommend this as a routine pre-exam preparation. Enteric-coated peppermint capsules (Pepogest, for example) are also helpful.

Can I take probiotics for stress-induced diarrhea?

One recent study of individuals complaining of stress showed that a probiotic containing a *Lactobacillus acidophilus* Rosell-52 and a *Bifidobacterium longum* decreased abdominal pain, and nausea and vomiting, but not psychological symptoms or sleep problems caused by stressful life events. Will probiotics work for you if you take them right before that big exam or meeting with your boss? Probably not. They likely need to be taken over a few weeks' time if they are going to help.

Before I had my gallbladder removed, I would have diarrhea whenever I ate fat, but now I have diarrhea every time I eat. What can I do?

This happens in 10 to 20 percent of people who've had their gallbladders removed. The gallbladder stores up bile and releases it into the duodenum when you eat. When your gallbladder is absent, the bile constantly bathes the small intestine. Most of the bile acids, contained in the bile, are absorbed in the ileum, the last part of the small intestine. However, if they are not absorbed, they travel into the colon, where they act as an irritant and cause diarrhea. Medications that bind bile are useful: these medications are cholestyramine (Questran), colestipol (Colestid) and colesevelam (Welchol).

Every time I eat, I get cramps and diarrhea. My doctor says that I have IBS with diarrhea. Could there be something in my diet causing a problem?

One of the most common food products that causes diarrhea (and gas) is milk-containing food or drink. The enzyme lactase, which breaks down the sugar, or lactose, contained in milk is reduced or lost as many people age or after an intestinal infection. The incidence is lowest in Caucasian adults (7–20 percent). It is high in other ethnic populations, with 50 percent of Hispanic adults, 65–75 percent of African and African-Americans adults, and 80–95 percent of Native American adults being lactose-intolerant. When the undigested lactose gets into the colon, the colonic bacteria break it down and produce hydrogen gas and fatty acids that cause diarrhea and often pain. Lactase-treated milk or ingestion of lactase in a supplement form may help. You can diagnose the condition by taking a breath test to see if you release hydrogen after lactose ingestion. However, I recommend that you just stay away from all milk-containing products for one week and see if you get better. Other substances that may cause diarrhea are sorbitol (in chewing gum) and fructose (in fruits).

I love to run for exercise and have competed in a number of marathons, but it's a struggle. About five miles into my runs, I start getting abdominal cramps and a terrible urge to find the nearest woods. I have not had any bleeding from my stomach or intestines, although some of my friends who run marathons have. What can I do to continue to compete?

This is a common complaint among runners—and curiously more common with running than with other sports. In fact, during intense running, as many as one-third of runners may have abdominal pain, and a number of these runners feel the same way you do—where's the nearest tree or toilet? The cause for the pain is more elusive. Rarely is there found an anatomical problem that can be "fixed." Mechanical tugging on the tendons and ligaments, trauma caused by part of the intestine rubbing against a muscle or a decrease in blood flow to the intestine are thought to be potential causes. It is possible that the release of hormones during running can stimulate contraction of the colon. In a triathlon it's usually the running part that is associated with pain. Up to nine in ten runners may have hidden blood in their stools after a long endurance race. Rarely does the bleeding become visible, but this can happen. The bleeding can be due to ulcerations in the stomach or decreased blood supply with ischemic colitis in the colon.

Try these strategies:

1. If you have a high-energy bar before or during the race, try practice runs with and without the supplement and see if the bar could be contributing to the symptoms.

2. Make sure that your training is gradual and you don't abruptly increase your distance.

3. If you do get the pain, bend forward, tighten your abdominal muscles, take a deep breath and breathe out slowly through pursed lips.

4. If these techniques don't work, I have found that some runners benefit by taking under the tongue an antispasmodic (hyoscyamine) either before the race or about a half hour to one hour before symptoms usually start. If you would like to try this approach, try it during a practice run first! This medication possibly could cause you to feel thirsty during your run. It also has to be obtained by prescription.

5. If you become anemic after your runs (without visible bleeding), then taking an over-the-counter H2 blocker (a medication that blocks the histamine receptor type 2 induced acid releases) or PPI proton pump inhibitor that blocks acid secretion may help. At this time, for visible bleeding from the rectum during running, there's not much we can do, and if it happens frequently (which is very rare), then you might have to cut back on running and try another sport.

I have a dirty secret that I can't share with anyone. Even though I'm only fifty years old, when I get diarrhea, I can't hold it. I've been up in the middle of the night far too often, washing sheets. I've had to dash into a bathroom while shopping, only to find I didn't make it. I now have to carry underwear with me and wear an adult diaper. It is mortifying. Am I the only one? Why does this happen? What can I do?

Leakage of stool is terribly embarrassing at any age, and I hear these questions from my patients all the time. It can strike at any age but occurs more often in the elderly. You are not alone. In Canada 3 percent of teenage girls experience fecal incontinence, and in the United States 3 percent of women aged twenty to twenty-nine, 10 percent of women aged forty to fifty-nine, 14 percent of women aged sixty to seventy and 22 percent of women over eighty years of age have this form of incontinence.

Does your stool come too fast due to diarrhea? Is your stool too liquid, or can you simply not feel the stool coming at all? These could be contributing to your fecal incontinence, defined as the leakage of stool out of the rectum. Fecal incontinence has many causes. It can be due to problems with (1) the muscles in the anus and pelvis, (2) sensation in the anus and (3) the ability of the rectum to store the stool. Inflammation in the rectum, poor muscle function, and abnormal nerve function or supply to the anus and rectum can be causes for this loss of control. (Some people might suspect it has to do with how many times a woman has given birth; although trauma to the anal canal during birth, such as a tear, can contribute, it's not a direct cause.) Other factors include relaxation of the musculature and connective tissue as a woman ages and illnesses such as diabetes, which can affect muscular or nerve function. Removal of hemorrhoids can sometimes result in stool incontinence, too.

A good digital evaluation of your rectum and anus should be done by your doctor. This should be done both while you are relaxed to test for resting tone of the internal anal sphincter and while you are trying to squeeze your anus (the way you would if you were trying to hold stool in!) to test your ability to tighten your external anal sphincter. There are other tests used to evaluate the problem if simple measures as described below do not work. Usually a gastroenterologist will order the special tests.

I often tell my patients to try Kegel exercises. They can sometimes strengthen the external anal sphincter, allowing you to hold the stool for longer periods of time. Squeeze your buttocks and anal muscles and think about pulling up toward your shoulders. This should cause the external anal sphincter to contract (unless there is a problem). To try to strengthen the sphincter, do three sets of ten squeezes two to three times per day.

Oftentimes, patients come to me thinking they're doing their exercises properly, but nothing is happening. In this case, I try biofeedback. During biofeedback a catheter is placed into the rectum and monitoring of the anal sphincter pressure is done. This allows you to see the pressure increasing on a monitor when you are actually tightening the muscle, and it allows you to make adjustments to your technique.

If your incontinence is due to diarrhea, then the diarrhea should be evaluated and treated. Constipation might also be to blame—one not-so-obvious problem that can cause diarrhea and possibly incontinence is extreme constipation. The liquid stool leaks around the hard stool and looks like diarrhea. This often results in a mistaken diagnosis of diarrhea and not constipation. Make sure to let your doctor know if you have been having a problem with constipation.

Finally, never hold it in. If you have the urge to go—go!

My hemorrhoids are killing me. They flare up if I get constipated and strain, or if I get a GI bug and have terrible diarrhea. What can I do to make them go away?

Hemorrhoids are dilated veins of the rectum. These can cause bleeding, itching, pain and projection out of the anus. There are two types: external and internal. The treatment of hemorrhoids depends on the type and severity of the symptoms. For mild symptoms, I recommend medical therapy. A diet with twenty to thirty grams of fiber might bulk up the stool and can help prevent hemorrhoidal bleeding, but not hemorrhoidal protrusion. Diarrhea should be treated if present, and don't strain! My tips:

1. Drink plenty of water.

2. Ask your doctor about fiber supplements, which might help to stop bleeding.

3. Mineral oil is also worth discussing with your doctor—it eases the passage of stool. It should not be used for long periods of time and should be taken with breakfast or lunch—not before going to bed.

4. Soaking in a tub of warm water often helps itching or irritation.

5. Local steroids, which are present in many over-the-counter and prescription medications, may help. Pain can be reduced by local analgesic compounds, like Preparation H, Analpram and Anusol. These products should be used for limited periods of time.

6. Balneol lotion can often help itching.

Bear in mind that sudden pain and firmness in the hemorrhoids may indicate a clot. These clots aren't dangerous, though the pain can be excruciating. Relief can be obtained by removing the clot if done within forty-eight hours. Otherwise if the external hemorrhoids aren't causing problems and haven't changed, leave them alone. Surgery can be extremely painful.

You can sometimes help relieve the pain of swollen, protruding internal hemorrhoids by pushing them back inside. If you continue to have symptoms with the internal hemorrhoids, or they are protruding with minimal or no straining, then rubber banding (a procedure in which the hemorrhoid is tied off at its base with rubber bands, cutting off the blood flow to the hemorrhoid and causing it to clot) or another technique, such as sclerotherapy (obliterating the opening of the vein and making it hard), should be the next step. If your hemorrhoid is unresponsive and truly needs to come out, a colorectal surgeon would do the surgery.

Celiac Disease

Diarrhea lasting for weeks, months and years is debilitating. Why does it happen? Some common causes are celiac disease, IBS with diarrhea and inflammatory bowel disease (which I'll talk about in Chapter 7).

Celiac disease, a condition involving gluten intolerance, has become increasingly common, possibly due to more consumption of wheat and gluten products or to bacterial changes in the gut. Still, many physicians think that it is a disease of the Irish, Welsh and British, because studies in the past pinpointed the highest incidence among these groups. This just isn't true. About 1 percent of Americans have celiac disease (also known as gluten-sensitive enteropathy, celiac sprue and nontropical sprue). It is likely that this figure is going to be higher in the future. Between 1995 and 2003, the incidence of celiac disease increased more than fourfold in one area of Minnesota, and this spike is indicative of celiac disease's prevalence in other areas.

Typical symptoms of celiac disease include fat malabsorption with diarrhea, weight loss, and signs of vitamin or nutrient deficiencies, anemia, and loss of bone. (Note: you don't have to have diarrhea with celiac disease. Some people get constipated instead.)

You said that celiac disease is often a reason for diarrhea. I've never heard of it—it sounds like a medieval curse!

Simply put, celiac disease is an intolerance to gluten—the gluten causes damage to the small intestine, especially near the stomach. You'll find gluten all over the place, in things like wheat, barley and rye—many of our everyday foods. Most breads, fried meats (since they're rubbed with flour), sauces, thickening agents, pancakes, and even some ice creams, contain gluten products. Labels are key to knowing if processed food contains gluten. Be careful. Just because a cereal box says it contains rice cereal, that doesn't mean it is free of wheat or gluten. Look for a label that says "gluten-free" to be sure!

With celiac disease, because of the damage to the lining of the small intestine, food products aren't properly absorbed. This results in diarrhea, loss of fat and other nutrients, and weight loss. Celiac disease often causes iron deficiency anemia that can't be explained by menstruation or other blood loss. But that's not all. Other symptoms include fatigue, osteoporosis, skin rashes, canker sores, liver disease, muscle and joint aches, headaches, asthma, depression and neurological problems. Celiac disease is strongly associated with other autoimmune diseases, such as type 1 diabetes, autoimmune thyroid disease, IgA (immunoglobulin A) deficiency and IgA-caused kidney disease, dermatitis herpetiformis (an itchy, blistering rash), Down syndrome and psoriasis.

Why does it happen?

In the United States anywhere from 1 in 80 to 1 in 140 people suffer from celiac disease. Almost always, the victim has a hereditary predisposition to the condition with the gene HLA DQ2 or HLA DQ8, which can be detected via blood tests. In first-degree relatives such as mother and child or sibling and sibling, the incidence is 1 in 22 people, while in second-degree relatives, such as a grandparent, aunt or uncle, or first cousin, the incidence is 1 in 39 people.

Although up to 40 percent of people in the United States are genetically predisposed, only about 5 percent of those people develop celiac disease. Why? A predisposition sets you up for the disease, but something else has to happen to cause it. Early infant exposure to cereal that contains barley, wheat or rye is definitely a factor. Gluten exposure in the first three months of life is also associated with over a fivefold increased risk of developing celiac disease. Breastfeeding infants at the time gluten is introduced into the diet may decrease their subsequent risk for developing celiac disease by over 50 percent. Recently the European Society for Pediatric Gastroenterology, Hepatology and Nutrition recommended introduction of gluten at an age older than four months but younger than seven months. Intestinal-cell damage due to a GI bug or GI tract surgery is another factor.

For sufferers, the gliadin protein (a portion of gluten) travels through the stomach and crosses the intestinal border, where it interacts with an important protein, called tissue transglutaminase, needed for the health of your gut. The tissue transglutaminase alters the gliadin protein, allowing it to bind tightly to cells, which then set off several immune reactions. The final result is damage to the intestine. Antibodies (proteins made by immune cells in the body to fight against bacteria, viruses or other substances) are produced against tissue transglutaminase. The measurement of these antibodies is the basis for diagnosing the disease. In infants, celiac disease is often suspected if they are losing weight, are having trouble gaining weight or have an extended bout of diarrhea.

I thought celiac disease was a childhood illness, but my friend was diagnosed in her forties and ate bread all her life. Is that typical?

We used to think that celiac disease showed up in childhood, before we became more aware of the variety of presentations and symptoms of this condition. In the past, celiac disease was typically diagnosed in children between the ages of four months and two years. Typically, celiac disease is now diagnosed when a person is in her forties. Sometimes celiac disease in children goes away, only to return in adulthood, in spite of eating a normal diet.

Can I get my blood tested for celiac disease?

People with celiac disease have higher than normal levels of certain auto-antibodies—proteins that react against the body's own cells or tissues—in their blood. To diagnose celiac disease, doctors will test blood for high levels of IgA anti-tissue transglutaminase antibodies (tTGA) or antiendomysial antibodies (EMA). If test results are negative but celiac disease is still suspected, additional blood tests may be needed. Tests for IgA tissue transglutaminase (tTG) and antiendomysial antibodies are the most sensitive and specific. If you have these antibodies, doctors will strongly suspect that you have celiac disease. All testing for antibodies should be done when you are eating gluten (bread, pasta, etc.), since the antibodies decrease and often return to normal after one to six months on a gluten-free diet. (It takes two to four weeks for the antibodies to reappear or increase after a person goes back on gluten after being on a gluten-free diet for some time.)

Although antibody testing is excellent, I still recommend that a diagnosis be confirmed by taking a sample of the lining of the small bowel (usually done at the time of an upper endoscopy in which a flexible tube with a light is inserted into the mouth and advanced through the esophagus and stomach into the duodenum, the first part of the small intestine) in everyone except people who have had a skin biopsy showing they have dermatitis herpetiformis (a blistering, very itchy rash), which is also caused by gluten toxicity.

Once you're diagnosed, you'll need to modify your diet. Following a gluten-free diet is a lifetime commitment, and I'll discuss it in more detail later on. Luckily, your tTG or antiendomysial antibodies can be used to track your disease. If the antibodies go up, it's possible that you're ingesting gluten (knowingly or unknowingly). If you're still experiencing diarrhea symptoms despite modifying your diet, you will need further evaluation. The first test should be an upper endoscopy to rule out refractory sprue (unresponsive to a gluten-free diet) or a tumor (such as lymphoma, a tumor of the lymphoid tissue in the gut). Sadly, although celiac disease is becoming more and more prevalent, it's not always diagnosed quickly. In my experience, this is due to two things: (1) many sufferers don't want to face the fact that something is amiss, so they put off going to the doctor; (2) other times, physicians don't hone in on this diagnosis right away, perhaps because it's still relatively new.

Such was the case with my friend Julie, a forty-year-old professor and mother of three. She'd suffered from a host of unusual physical symptoms, coupled with extreme fatigue and depression. One day she mentioned her problems to me, and I suggested she ask her doctor to test for celiac disease. She wanted to share her story.

One thing comes to mind when I think of the rather long period of time during which I was sick with celiac disease and undiagnosed: overwhelming, chronic, debilitating fatigue. The onset was insidious and gradual, making the fatigue seem "normal" somehow. I began to think that feeling exhausted, snappish and at the end of my rope all the time was just the way things were, a by-product of having a busy life and getting older. My doctor would say, "Of course you're tired all the time. You have three children, a demanding job and some major stressors in your life." I assumed that all my friends and colleagues, at least the working moms, dragged themselves through their days the same way I did. It was not until the healing that came with the gluten-free diet that I even began to remember what a normal energy level felt like.

It's been about a year and a half since my diagnosis, and while I have had "down" days since then, I have not experienced anything like the emotional gloom that was my almost constant companion for some years before. The feeling was so qualitatively different from anything that I have had since, I assume that it was a depression very rooted in the physiology of the disease. Given the nutritional deficiencies, the "toxic" feeling in the head and body that I now recognize as gluten, the emotional issues that come with feeling chronically unwell—all compounded by fatigue—it would be a miracle not to be depressed!

The question of exactly how long I had celiac disease before I was diagnosed interests me, but it's very difficult to know. I have read about the suspected triggers, in addition to gluten, that might throw a genetically susceptible person into an active disease state, for example, childbirth, surgery, various infections and severe emotional stress. Yes, yes, yes and yes. About ten years before diagnosis, I had my third child, underwent tubal ligation and was stressed. In my mind, this clustering of events seems like my "trigger" and fits with the timing of when at least a subset of symptoms really began to ramp up.

Until about eight years before my diagnosis, anemia, some depression and a bit of winter asthma had been my only health issues. I was hardly ever in my primary care doctor's office. Over the eight years before my diagnosis, it was a long string of health problems, one right after another. Some, I have learned, are directly related to celiac disease. In the time approaching my diagnosis, I felt that I was practically living at my doctor's office and assumed that she had started to see me as a hypochondriac. From that point eight years before diagnosis forward, the list at varying times included fatigue, severe and protracted headaches, anemia unresponsive to oral (iron) supplementation, worsening depression, hair loss, dizziness, very low adrenal hormones (DHEA and DHEA sulfate) and testosterone, osteopenia (loss of calcium in the bones), pronounced weight gain, a large ovarian cyst and uterine problems leading to hysterectomy.

At least a year before diagnosis I developed gastrointestinal problems, mostly abdominal bloating and pain. There was some diarrhea, but nothing dramatic until toward the end. These symptoms, too, had a very gradual onset and seemed to just become my new "normal." I attributed them at least in part to the very high levels of iron that I was then taking orally. At the same time, I had pelvic problems, which confused the source of the discomfort. I remember my doctor saying something to the effect that most women could distinguish between pelvic and GI pain. It felt like a chastisement then, but I can see now that she was frustrated by her inability to make a diagnosis and help me.

Based on the pelvic and abdominal pain, I had a CT scan, which helped to diagnose adenomyosis (a condition where the cells lining the uterus grow into the muscles of the walls of the uterus). I remember that the radiologist put some disclaimer on the report about fluid in the bowel loops (or something like that), suggestive of malabsorption, and recommended ruling out celiac disease, which I had never heard of at that time. When I asked my doctor about this, she said that it was the new, trendy thing and was recently showing up on a huge number of radiology reports. She told me not to worry about it, that it was just an "FYI" thing for the radiology department. That was more than a year before my diagnosis! An easy blood test at that point could have at least spared me that last year of ill health, which was definitely the worst.

The uterine problem caused pelvic pain, but neither heavy nor prolonged bleeding. However, I still needed the hysterectomy and it was this, presumably unrelated to celiac disease, that probably helped the most in getting to a diagnosis. The gynecological surgeon pointed out my very anemic state and made a very specific recommendation in the chart that this be worked up and followed closely after the hysterectomy. My iron levels didn't bounce back even after months of oral supplementation. I continued to feel worse—pain and bloating, tremendous fatigue, the works—and it became more difficult to dismiss my symptoms. As the weeks and months progressed, I felt that I really couldn't function much longer in that state. One day I was driving at sixty miles per hour on the highway with my young daughter in the backseat when I had a dizzy spell that verged on blacking out. I had to pull over and just sit with my head down for what seemed like forever.

    Back, again and again, to my doctor. Even then, the celiac disease bells were not ringing for her. Instead, she assumed that I continued to be anemic because of a GI bleed and scheduled me for a colonoscopy to find it. At about this time, in tears, I called Dr. Wolf. She explained to me that if you are chronically anemic, it could be because you are losing iron by bleeding or because you're not absorbing it well to begin with. She gave me the specifics on the blood test to request for celiac disease. Without this intervention, I have no idea how many more years I might have floundered without being tested, and I am intensely grateful for her input.

    When I told my doctor about this conversation and asked her for the blood test for celiac antibodies, she absently flipped through my now huge medical record and asked, "Didn't we do that?" No, we hadn't, and she continued to talk about it as a very low probability situation. Of course, the test came back screamingly positive. I got connected with an excellent celiac specialty center; got an intestinal biopsy, which confirmed the diagnosis; started the gluten-free diet; and the rest is more or less history. I think my primary care doctor was probably trained to think of celiac disease as very rare and to look only for the classic picture, a condition seen mostly in children with GI distress and wasting as the cardinal features.

    Since my diagnosis, the improvements have been wonderful, but interrupted. I developed Hashimoto's thyroiditis, a common autoimmune condition (inflammation of the thyroid gland) that is even more common for people with celiac disease. It took a while to get sorted out and on the right dose of thyroid hormone to feel better. It is an ongoing process to get it right with the gluten-free diet now that I am, of course, absorbing everything. But everything is on the right track now, and I look forward to continuing to take back my life.

Do you have to starve yourself if you have celiac disease? It seems incredibly limiting.

Trust me, it is possible to live a normal life with celiac disease. In fact, many restaurants are even getting in on the act, with gluten-free dishes and menus. If you dine in a restaurant that does not have these choices, discuss what you can and can't eat with the waiter. Many of my patients find it helpful to carry a card listing what they can and cannot eat. Slip it to the chef to guarantee a gluten-free meal. Remember: You can't even eat food cooked on the same grill where breaded meat or fish is cooked, since there can be contamination. But you should be able to enjoy eating despite your disease.

Indeed, managing your diet is a long-term challenge. Katie was a twenty-one-year-old college student when she came to see me during her junior year of college. Part Hispanic, Irish and Native American, she loved to travel and explore new places—but her symptoms were beginning to squelch her adventurous spirit.

She'd taken a year off and lived in Australia the year between high school and college. The food was great, she told me—and much to her dismay, she gained twenty pounds while abroad. When she returned, though, she quickly shed the twenty pounds and then began to lose even more weight. Her weight loss was accompanied by bloating, abdominal cramps, fatigue and diarrhea. A blood test was positive for celiac disease, and she underwent an upper endoscopy with a sampling of the duodenum, the first part of the small intestine, which confirmed the celiac disease diagnosis. She was put on a gluten-free diet and started school.

She should have been pulling all-nighters in the library or trying to decide which party to go to next. Instead, she was in the bathroom. For the first three to four months of school, she had diarrhea at about 10:00 a.m. every day, smack in the middle of her first class. As soon as class was over, she had to run back to her dormitory and dash to the bathroom, always worried that she might not make it. She also felt nauseated throughout the day.

She initially lost more weight on her gluten-free diet, which was very limiting—no late-night pizza runs for Katie. She usually ate nothing but brown rice pasta, sometimes switching to salad for variety. The summer after her freshman year she traveled a bit and developed a parasitic infection, which caused even more cramps and diarrhea. She was treated successfully with antibiotics.

Sophomore year, as she continued with her gluten-free diet, things improved. But during her junior year she deteriorated. She'd wake up in the morning with cramps in her right lower abdomen, which were worse after eating. Katie continued to get sicker; she began getting diarrhea whenever she ate, coupled with crippling stomach pain.

Not surprisingly, she completely lost her appetite. She visited Health Services on campus and was told there was nothing that could be done for her. She postponed exams and generally felt miserable, going to the bathroom up to four times per day. Still, she continued to eat in the school dining hall, trying to pick foods that she thought were gluten-free.

It was at this point that Katie came to see me. Her main complaint was cramping and off-and-on diarrhea; she also told me that asthma symptoms flared up whenever her abdominal pain started up. I considered her other attributes: she didn't smoke or drink; her weight was normal for her height. She did have a relative with celiac disease, which is common.

Because she had recently traveled to Africa, I first had to rule out any hidden infections. Katie's stool tests showed no infection or parasites. On the other hand, her blood IgA tissue transglutaminase antibody was mildly elevated, suggesting that her celiac disease was not under great control and was causing the diarrhea.

I suspected that Katie was somehow unintentionally getting gluten into her diet. Over the next two months she reported that she was trying to be better about her eating habits and making sure that she did not accidentally eat gluten, with good results. Her diarrhea was less frequent, although every so often it would resurface for about a week.

As a college student, Katie found it difficult to stick to a gluten-free diet on campus when her choices were so limited. She had had minimal success when she tried to work with Dining Services initially. After much persistence, she was able to enlist Health Services to join her as an active partner in advocating for healthy gluten-free options in the dining halls. Once Health Services was involved, more effort was made by Dining Services to provide students more healthy choices that were gluten-free. Other students, hearing about her efforts, revealed to Katie that they, too, had celiac disease and were greatly appreciative of her effort to improve food choices.

Essentially, Katie helped push for more salads, fruit and safely cooked meat. In addition, Katie went to a nutritionist who helped her understand which foods were gluten-free. The nutritionist went to the dining hall with Katie on a food tour and pointed out gluten-free foods. With the nutritionist's help, Katie was able to make better food choices. Her diarrhea totally disappeared, which in turn lowered her stress levels. She started multivitamins and calcium with vitamin D to make sure she was getting enough nutrients. With the improvement of her symptoms, her weight increased another five pounds. Katie is now a college graduate, happily living without diarrhea.

Katie's challenges were several. First and foremost was how to stick to a gluten-free diet while eating institutional food. If she didn't stick to a good diet, she would have abdominal pain and diarrhea and sometimes vomiting. The choices at school were limited, and Katie was not able to make healthy choices on her own. Furthermore, it required teamwork between Health Services and the dining facility to put together healthy choices for Katie and other students suffering from the same condition. Katie also had to cope with the weight fluctuation that came with her normal absorption of nutrients. Many people have to cut down on their calories when they start to better digest and absorb the nutrients and calories in their food.

What are the foods that are safe to eat for Katie and others like her with celiac disease? How could she make mistakes in the dining hall?

One reason that Katie was making bad choices is gluten's stealth nature: it can lurk in condiments, prepackaged foods and many food products, such as seasonings, colorings, flavorings, sauces, creamed foods, canned soups and salad dressings. The fact of the matter is, you need to be vigilant in finding out which foods contain which ingredients.

Aside from that, remember to avoid wheat (including spelt), rye and barley. Safe foods are corn, rice, potatoes, soybeans, tapioca, buckwheat, millet, amaranth, quinoa, arrowroot and carob. Meats, fish, fruits and vegetables are all fine, as long as there are no additives. (Flip to Chapter 9 for a list of ingredients that are safe to eat and those to avoid if you have celiac disease.) Interestingly, lipsticks, lip balm, Play-Doh and postage stamps often have gluten in them. So do harmless-seeming things like baby powder and suntan lotion. Be sure to read labels carefully to avoid accidentally introducing gluten into your diet. It takes extra work, but it's worth it!

Now that I'm following a gluten-free diet, do I have to worry about any long-term problems?

The nutritional deficiencies that occur with celiac disease can cause problems in the future. Lack of vitamin D and poor calcium absorption can result in bone loss. Adequate vitamin D levels (measured by blood tests), with supplemental vitamin D as necessary, and bone density evaluation and treatment if you are osteopenic (have a slightly low bone mineral density) or osteoporotic (have a low bone mineral density, putting you at increased risk for a fracture) are important. Other vitamins can also be lacking. Vitamin B_{12} deficiency can cause numbness and tingling. Zinc deficiency can cause brittle hair and skin rashes. You might end up having to take a multivitamin supplement. Be alert for the other conditions discussed above that are increased in people with celiac disease. There is a slight increased risk of lymphoma (cancer of the lymphocytes) that decreases with time after diagnosis of the celiac disease.

WHAT YOU NEED TO KNOW:

1. Food poisoning is common. Wash vegetables and fruit carefully. Cook meat thoroughly. If you like your meat less than well-done, you are increasing your risk for food poisoning.

2. Pepto-Bismol, Imodium and probiotics may be helpful if you have food poisoning. Do not use Imodium if you have blood in your stool.

3. Call your doctor if you have blood in your stool.

4. Pepto-Bismol, certain antibiotics and probably probiotics are useful to prevent traveler's diarrhea.

5. Celiac disease is a condition in which the body has a reaction against gluten, a substance contained in wheat (including spelt), barley and rye. It is treated by elimination of all gluten in the diet.

6. You can have celiac disease without diarrhea. If you have unexplained anemia, bloating and gas, or deficiency in a nutrient, be sure that you are tested for celiac disease.

7. Celiac disease can be diagnosed in most people by a blood test, although most doctors like to confirm the test results by doing an upper endoscopy to obtain a small piece of tissue from the duodenum, to be examined under a microscope.

Chapter 5
When You Just Can't Go: Constipation

"I wish that being famous helped
prevent me from being constipated."
—**Marvin Gaye**

T his chapter focuses on the other side of the coin: constipation. Moving your bowels should be easy, effortless and painless, if not exactly the pinnacle of fun. But for too many of us, this just isn't the case. In fact, between 12 to 19 percent of American women deal with chronic constipation. And as we age, the number creeps up to one in three women over sixty-five complaining of constipation. These women have a difficult time eliminating stool, have incomplete bowel movements or go less frequently than three times per week.

Why? Well, many people with constipation might not be eating right—not getting enough water or fiber and devouring too much junk food. (In fact, a typical American diet is poor in fiber; a healthy diet should contain twenty-five to thirty-five grams of fiber per day.) When I see a constipated patient, my first goal is to up his or her consumption of fiber. But this is just the first step. Often constipation is due to more than just poor fiber intake. Many medications cause constipation. The medications for heart disease, depression and pain are common ones. Some examples of medications that commonly have constipation as a side effect are:

Prescription drugs

Examples shown in parentheses

- Pain medications, particularly those containing opiates (morphine, oxycodone, codeine)
- Muscle relaxants (cyclobenzaprine hydrochloride)
- Antispasmodics (dicyclomine, hyoscyamine)
- Antidepressants (tricyclic antidepressants—amitryptyline, desipramine, imipramine; other antidepressants—sertraline hydrochloride)
- Anti-Parkinson's drugs (amantadine hydrochloride)
- Blood pressure medications (beta-blockers and calcium channel blockers) (propranolol hydrochloride; verapamil hydrochloride)
- Diuretics (Furosemide)
- Anticonvulsants (phenytoin)
- Antipsychotic drugs (ziprasidone hydrochloride, haloperidol, chlorpromazine)
- Antihistamines (dihydramine, loratadine, cetirizine hydrochloride)

Nonprescription drugs

- Antacids that contain aluminum and calcium
- Iron supplements (ferrous sulfate, ferrous gluconate)
- Antihistamines (see above)
- Antidiarrheal agents (Loperamide, Pepto-Bismol)
- Calcium supplements (calcium carbonate, calcium citrate)
- Nonsteroidal anti-inflammatory agents (ibuprofen, naproxen)

Constipation has many causes. Some people simply have slow movement of stool through the colon. In others, the muscles and anus don't work as efficiently, and the stool just doesn't come out well. Systemic conditions can also cause constipation. These include neurological disorders like multiple sclerosis, spinal cord disease, Parkinson's disease, diabetes mellitus, anorexia nervosa, low thyroid function (hypothyroidism) and other endocrine abnormalities. Irritable bowel syndrome is a common cause of constipation, but constipation does not have to be IBS.

How do I know if I'm constipated?

Do you have fewer than three stools per week, or has the frequency of your bowel movements decreased from your normal number? Do they come out in hard pieces? Does it feel like it's hard to completely eliminate your stool, or that you're straining to get it out? If so, chances are you're constipated.

The periodic constipation that we all get at some time or other may be due to hormonal fluctuations, travel, inactivity, poor hydration, or poor or changed diet. Often this type of constipation will respond to an increase in fluid intake and fiber intake (by diet or supplement) and an increase in vigorous exercise. A fiber supplement like psyllium or guar, or a stool facilitator like MiraLAX or GlycoLax from your local pharmacy, may be helpful.

However, if the stool just refuses to budge, or if you experience blood, pain or bloating, consult your doctor. Other warning signs that should push you to see your doctor are loss of appetite, anemia (low red blood cell count), unexplained weight loss, a family or personal history of inflammatory bowel disease, or a family history of colorectal cancer. (See Chapter 3.)

There's also chronic constipation. (See the box.) In this case, you've been constipated for three months or more. There are typically three root causes: The most frequent type (accounting for about 60 percent of constipation) includes those people with IBS with constipation. Up to 28 percent of women have a problem with defecation—in other words, getting the stool out of the rectum. Up to 13 percent of people have slow passage of the stool through the colon. We call that "slow transit" constipation.

The clinical definition of chronic constipation is having any two of the following symptoms for at least twelve weeks:

- Straining during bowel movements at least 25 percent of the time
- Lumpy or hard stool at least 25 percent of the time
- Sensation of incomplete evacuation at least 25 percent of the time
- Sensation of anorectal blockage/obstruction (the stool is down low, just above the anus, but won't come out) at least 25 percent of the time
- Use of a finger or hand to help the stool come out at least 25 percent of the time
- Fewer than three bowel movements per week

One of my patients, a fifty-two-year-old mother of two named Arianne, endured a long bout of irritable bowel syndrome with constipation. On the surface, you'd never suspect that she had any problems at all: Meticulously dressed, with perfectly coiffed blond hair and plenty of jewelry, Arianne was a witty, sophisticated woman who looked like she had her life supremely under control. She sat on several charity committees and moved in the upper echelons of society. But in actuality, in between fancy dinners and galas, she was running for the bathroom and stockpiling laxatives.

Arianne came to me because she was able to go to the bathroom only once or twice per week. She'd suffered from ovarian cancer years before; in fact, one of her symptoms at that time was abdominal pain and the constant need to go to the bathroom. Like my patient Elizabeth, whom we met in an earlier chapter, she was an expert at scouting out quality public bathrooms for when she had the diarrhea, as she had no time to spare when the urge hit. After successful cancer surgery, she flew to the opposite end of the spectrum—severe constipation. Her case is representative of some of the most frustrating issues I've faced.

> For the first forty-nine years of my life, I never gave much thought to my bowels. Then I developed severe diarrhea and was diagnosed with IBS. None of the medicines helped. I was told that I was one of those patients who "doesn't react to medication." After one and a half years of planning my life around my stomach, writing my own personal reviews of metropolitan restrooms (not only for cleanliness, but for ease of access too!), being in constant pain and bloated, I took charge of my life. I pressured my doctor for a vaginal ultrasound. My stomach problems were not IBS. I had ovarian cancer.

Arianne's ovarian cancer was treated with a total abdominal hysterectomy (the removal of her uterus), plus the removal of both fallopian tubes and ovaries. The abdominal pain and other GI symptoms that she had before the surgery disappeared. However, fifteen months later, the pain came back and her abdomen became distended. Because of the fear that she might have had a recurrence of the ovarian cancer, an abdominal CT scan was done. This showed no tumor, only a colon filled with stool. After a good dose of laxatives, she cleaned out her bowels, and the stomach distension decreased. A colonoscopy reassured Arianne that colon cancer was not causing her problems. Diverticulosis (a condition in which there are outpouchings of the colon) was present, but it was asymptomatic.

After surgery for the ovarian cancer, I thought I was done with IBS problems. I was so wrong. I began to suffer from extreme constipation.

I went to several doctors. I left my first doctor due to how he acted when he found out he had misdiagnosed me and had overlooked my ovarian cancer. Then I went to someone a friend recommended. After she said that she would normally recommend removing my colon for diverticulosis but in my case wouldn't recommend it, I asked quite innocently, "Why not?" I was told that in most cases ovarian cancer patients don't live that long, anyway, so why bother fixing my colon if I was going to die soon? I flew out of that office with my clothes barely on.

At the time of my first meeting with Arianne, she was having only one or two bowel movements per week. She felt pressure in her rectum; she told me it felt like her stool would get "stuck" and wouldn't come out. She also felt pain and fullness. Before her hysterectomy surgery, she would rush to the bathroom immediately after she ate in order to have a bowel movement. Since surgery, she had suffered from cramps that made her feel like she needed to go, but nothing would come out, despite a healthy intake of fiber. She ate two apples per day, drank a lot of water, exercised, took two calcium polycarbophil (FiberCon) tablets per day, took a capful of MiraLAX (polyethylene glycol 3350) daily and ate a lot of fruit. And yet she still had only one or two bowel movements per week.

Initially, I felt that Arianne had trouble getting her stool out of the rectum and perhaps had suffered some damage to the pelvic nerves during surgery. I started her on flaxseed, which helped soften her stools, but she still had a hard time going to the bathroom.

Simple treatment with fiber had failed. Because of coincident persistent abdominal pain and cramps, she was felt to have irritable bowel syndrome with constipation, but I questioned whether something else was wrong. I did some simple tests to see if she had slow movement through her colon or a problem with the functioning of her pelvic muscles needed to evacuate the stool or a poorly relaxing anus.

Testing showed that she had poor movement in her colon. When she swallowed a sitz-marker capsule that contains twenty-four little rings visible on an X-ray, most remained on the left side of the colon at five days, even though they should have been evacuated. However, by seven days they were almost gone, showing that the motility was only slightly slow. She had other problems, too, that were found with testing: she didn't feel the stool when it was in her anus, her anal sphincter didn't fully relax when the stool was trying to come out, the pelvic muscles didn't work very well and she had an outpouching of the rectum toward the vagina (rectocoele) when she tried to eliminate stool.

Each one of these problems required a different type of treatment. First, in an attempt to get the stool moving through the colon better, I tried various laxatives and bowel stimulants.

Through proper testing and Dr. Wolf's taking the time to learn all about my medical history, I was diagnosed with IBS with constipation. My constipation has been so severe that frequently I forget that the simple bodily function of the bowel movement even exists. On the rare occasion that my body decides to function, a feeling of urgency hits me. Then I have to revert back to my old list of quality public bathrooms.

Arianne's Bathroom-Scoping Tips:

Stuck in the mall and don't have time to run to a department store restroom? Those in Pottery Barn and Talbots have easy access (no keys needed) and are usually clean. Most Gap stores have restrooms, too. Staples, Home Depot and Lowe's usually have fairly clean restrooms but entail a long walk through the stores, so time and speed are essential. Near a hotel? Just walk in with your head high; no one stops you. If you're on the highway, don't get off unless you see an advertised gas station or restaurant (the home of emergency bathrooms!). Blundering around off an exit could mean a five- to seven-mile ride to a restroom.

I have tried every prescription medicine for IBS (see Chapter 3) along with other drugs that cause diarrhea. I've tried a lot of over-the-counter medication, too, especially cherry-flavored magnesium citrate, used to stimulate my bowels. I have rated the taste of different drugstore brands of magnesium citrate: Rite Aid's brand has the best flavor, then CVS, and then Walgreens. I used a full bottle of magnesium citrate for weekly relief. It meant taking it on Friday or Saturday evening and committing not to go out the next day. Not only did this kill a half a day on a weekend for me, but it also ended my desire to drink black cherry soda, once a favorite of mine. But the magnesium citrate just wasn't enough.

I tried holistic treatment; I then spent a considerable amount of money trying different supplements. The theory was that the supplements would balance my system, and then my bowels would work properly. I did this for months, and again, no results.

The next two years were difficult with regards to the constipation. Arianne began alternating different laxatives, as they would work a little and then stop. Then, one day, things were desperate. She ingested a whole colonoscopy preparation, drinking a full gallon of GoLYTELY, but it lingered in her belly— nothing came out. She tried more laxatives and still nothing. Scared that her bowel was blocked, she went to the emergency room, where an evaluation just showed fluid and stool retention, but no blockage.

I discovered the severity of my condition when a gallon of GoLYTELY did nothing for me. I was sent home from the emergency room in the same uncomfortable state in which I arrived, as no one knew how to help me find relief. A few days later I was fortunate to learn about colon hydrotherapy (for an explanation of this see page 98). I had several treatments and finally had some relief. In conjunction with it, I also had scar massage therapy for abdominal adhesions. This was somewhat painful, but I felt I had to have it done to rule out scar tissue as a complication to my bowel problems. I have been having the colon hydrotherapy on a regular basis for over two and a half years.

Colonic treatment is the only thing that gives me relief from both pain and fullness. However, I do wonder what's next for me. I can't picture doing this for the rest of my life. Sometimes I worry about the treatment center closing, especially since it's the only one in the metropolitan area. When it was closed for several weeks this fall, I tried in vain to find another location. I reverted back to my old ways of trying different types of treatment in succession (magnesium citrate, followed by suppositories, followed by laxatives, and then suppositories again, all in combination with fiber supplements). Nothing worked, and I was counting the days until the colonic center opened. I've become close with my therapist, and we spend each session gabbing away. I have watched her grow in her own personal life and wish only the best for her, yet part of me wonders what I would do without her. Some people become close with their hairstylist or manicurist, but for me, it's my colonic therapist.

Whenever I get frustrated by my current IBS situation, I remind myself that I am really fortunate and that I could be dealing with a lot worse challenges in my life. My sense of humor has always served me well in getting through the rough patches in life. So when I'm backed up and feeling lousy, I just announce, "I'm going to get plunged!"

Now, nine years after her ovarian cancer following a brief period of diarrhea, Arianne has a bowel movement daily, taking only probiotics and monthly colonics.

Arianne tried many treatments for her constipation. How do you decide what to use for constipation?

My general guide for treating constipation is to start with the product or procedure with the smallest effect on the bowel or body. If that fails, I move on to another type of product. First I assess the diet to make sure that the person is consuming enough calories and fiber, and I ensure she doesn't suffer from an eating disorder. If a person is getting less than thirty-five grams of fiber, then I recommend more.

Fiber is material contained in plants that can't be digested by the small intestine. A major fraction of fiber is carbohydrate components. Lignin is the main non-carbohydrate component of fiber. *Soluble* fiber (fiber able to be dissolved in water) passes relatively unchanged through the small intestine, where there is a limited number and type of bacteria, into the colon where it encounters a large number of many kinds of bacteria. The bacteria ferment the fiber, producing gases and short-chain fatty acids, which stimulate the bowels and hold on to water, bulking up the stool. Soluble fiber is concentrated in legumes (dried beans and peas), oats, barley and soy. *Insoluble* fiber passes through the colon relatively unchanged. This fiber can absorb four to five times its weight in water. Insoluble fiber is concentrated in wheat bran, most vegetables and fruit roughage. The fiber content in food listed in reports varies and this is likely due to the method used to estimate the fiber source.

The fiber in foods varies in its effectiveness in bulking up the stool. Studies suggest that stool weight is most effectively increased by wheat bran, followed by fruits and vegetables, oats, corn and soybeans. Cereals are a good source of wheat bran; you could also just add a couple of tablespoons of wheat bran to your favorite cereal, yogurt or applesauce. Of course, if you have celiac disease and constipation, be very careful in your culinary selections. The wheat products are out! However, information on fiber supplementation is poor regarding the effect of these foods on the stools of people with chronic constipation. (See the chart for common foods with the most fiber.) I have found that prunes are a good addition to a diet for helping constipation. You can eat one to five prunes daily. Steeping the prunes in hot water and eating the prunes as well as drinking the liquid may be even better (although this has not been proven). Rhubarb is another excellent choice for those seeking to add more fiber to their diet. Note that rhubarb has a laxative effect, as well as a fiber effect.

TABLE 5-1: FIBER CONTENT OF SELECTED FOODS

Food	Amount	Total Fiber (grams)	Soluble Fiber (grams)	Insoluble Fiber (grams)
Cereal				
Fiber One	½ cup	14.0	1	13.0
All-Bran Original	⅔ cup	13.0	1	12.0
Raisin Bran	1 cup	8.0	1.2	7.2
Shredded Wheat	2 biscuits	6.0	trace	6.0
Oatmeal (dry)	½ cup	4.0	1.9	2.1
Raw Vegetables				
Carrots	1 long	2.3	1.1	2.2
	6 baby	2.8	1.4	1.4
Celery	1 cup chopped	1.7	0.7	1.0
Onion	½ cup chopped	1.7	0.9	0.8
Pepper (green)	1 cup chopped	1.7	0.7	1.0
Cabbage (red)	1 cup	1.5	0.6	0.9
Tomato	1 medium	1.0	0	1.0
Lettuce (iceberg)	1 cup	0.5–0.8	0.1	0.4–0.7
Cooked Vegetables				
Turnip	½ cup	4.8	1.7	3.1
Rhubarb	½ cup	2.9	0.3	2.6

TABLE 5-1: FIBER CONTENT OF SELECTED FOODS

Food	Amount	Total Fiber (grams)	Soluble Fiber (grams)	Insoluble Fiber (grams)
Peas	½ cup	4.3	1.2	3.1
Okra (frozen)	½ cup	4.1	1.0	3.1
Potato w/skin	1 medium	2.9	1.2	1.7
Brussels sprouts	½ cup	3.8	2.0	1.8
Asparagus	½ cup	2.8	1.7	1.1
Broccoli	½ cup	2.4	1.2	1.2
Cauliflower	½ cup	1.7	0.4	1.3
Green beans	½ cup	1.9	0.8	1.1
Fruits				
Figs (dried)	1½	3.0	1.4	1.6
Apricots	6 dried halves, 2 fresh w/skin	1.7	0.9	0.8
Raspberries	½ cup	4.2	0.4	3.8
Mango	½ small	2.9	1.7	1.2
Orange	3-inch diameter	4.4	2.6	1.8
Pear w/skin	3-inch diameter	4.0	2.2	1.8
Apple w/skin	3-inch diameter	5.7	1.5	4.2
Strawberries	½ cup	1.9	0.5	1.4
Banana	1 small	2.2	0.6	1.6
Prunes (dried)	5 medium	2.8	1.7	1.1
Raisins	2 tablespoons	0.4	0.2	0.2
Cooked Legumes				
Kidney beans	½ cup	5.8	2.9	2.9
Navy beans	½ cup	5.8	1.9	3.9
Black beans	½ cup	6.1	2.4	3.7

TABLE 5-1: FIBER CONTENT OF SELECTED FOODS

Food	Amount	Total Fiber (grams)	Soluble Fiber (grams)	Insoluble Fiber (grams)
Pinto beans	½ cup	6.1	1.4	4.7
Lentils	½ cup	8.0	1.0	7.0
Grains				
Wheat bran	½ cup	12.3	1.0	11.3
Wheat germ	3 tablespoons	3.9	0.7	3.2
Rice (brown, cooked)	½ cup	1.7	0.1	1.6
Rice (white, cooked)	½ cup	0.2–0.8	0	0.2–0.8
Millet (cooked)	½ cup	3.3	0.6	2.7
Nuts and Seeds				
Flaxseeds	1 tablespoon	3.3	1.1	2.2
Almonds	6 whole	0.6	0.1	0.5
Walnuts	2 whole	0.3	0.1	0.1

How you choose to increase your fiber will depend on cost, effectiveness and the tolerability of gas as a possible side effect. It's usually cheaper to increase the fiber in the food rather than to add a supplement. Calorie intake is also a consideration when you add fiber. Sometimes a supplement will provide fewer calories than a large amount of cereal, for example.

Table 5-2 shows common fiber supplements and their total fiber content. Many of the store chains have their own generic brand of fiber compounds. Numerous preparations combining different types of fiber, and often including other substances, are available over the counter. Considerations in your selection of supplements are (1) cost (shop around); (2) tolerability, i.e., texture, taste and causation of gas and bloating; and (3) effectiveness (soluble versus insoluble fiber).

Make sure that you drink at least one eight-ounce glass of water or juice with each dose of supplement.

TABLE 5-2: OVER-THE-COUNTER FIBER PREPARATIONS

Preparation	Dose	Total Fiber	Soluble Fiber	Comment
Guar powder caplets/tablets	1–3 times/day 1 teaspoon 3 cap/tab	3 grams 3 grams	100%	Powder completely dissolves; gluten-free; drink large amount of fluid with it
Wheat Dextrin Benefiber Clear SF	2 teaspoons	3 grams	100%	Less than 20 ppm gluten
Inulin FiberChoice Fibersure	2 tablets 3 tablespoons	4 grams 15 grams	100%	Considered prebiotic
Calcium polycarbophil FiberCon Fiber-Lax Equalactin Mitrolan	1–3 times/day 2 caplets	1 gram	100%	Does not ferment in the colon; 244 mg calcium
Methylcellulose Citrucel	1 scoop 2 caplets	2 grams 1 gram	100%	Does not ferment in the colon; gluten-free
Pectin Apple pectin Grapefruit pectin	1–2.8 grams	0.8–2.2 grams	100%	
Psyllium Metamucil Fiberall Hydrocil Konsyl Perdiem Serutan	6 capsules 1 tablespoon 1 teaspoon	3 grams 3 grams 6 grams	67% 2 grams	
Flaxseed	1 tablespoon	3.3 grams	1.1 grams	Available whole or ground

In IBS constipation, the only fiber that has been shown to work in multiple studies is psyllium (Metamucil, Konsyl). One study showed the effectiveness of calcium polypcarbophil. However, many other fibers have worked in my experience. I often recommend a combination of organic flaxseed (whole or ground flaxseed is available at many markets) with a soluble fiber supplement like guar.

Fiber is most likely to work in those with constipation despite a normal transit through the bowel. In this case the fiber will help modify stool consistency. People who have slow movement through the colon (caused by a motility problem, drugs, a neuromuscular problem or poor evacuation from the rectum) are unlikely to be helped by fiber alone.

What if fiber fails? What is the next step?

1. Stool softeners are used in some people but have not really been shown to help constipation. They do draw water into the stool and make it less painful to have a bowel movement.

2. Lubricant laxatives (mineral oil) allow the stool to slide out of the rectum. This is helpful if someone has a fissure or hemorrhoids. Mineral oil is given as one to two tablespoons with breakfast or lunch. Note: It's dangerous to use if you have gastroesophageal reflux disease or to lie down after taking it. Mineral oil can decrease the absorption of Vitamin K and is dangerous in someone taking blood thinners. It can also interfere with the absorption of nutrients and therefore should be used only for a short period of time.

3. Osmotically active laxatives (lactulose) aren't absorbed in the small intestine and are fermented in the colon. Consequently, they may result in gas and discomfort when used. I suggest one to two tablespoons up to three times per day. If you develop gas or discomfort it should resolve shortly after stopping the medication.

4. Polyethylene glycol 3350 (MiraLAX, GlycoLax) stays in the bowel lumen (opening of the hollow-tube-like intestines) and draws fluid into the lumen, bulking up the stool. The usual dose is seventeen grams, but occasionally thirty-four grams are needed. My experience suggests that taking the thirty-four grams at one time is better than seventeen grams twice a day, but this has not been tested in any trial.

5. Saline laxatives draw water into the stool and increase intestinal contractions. Typical laxatives in this category are milk of magnesia (two to three

tablespoons/dose) and magnesium citrate (one half to one bottle/dose). You should not use these magnesium-containing laxatives if you have kidney disease.

6. Stimulant laxatives work by increasing movement in the intestines. They should not be used for a long period of time unless you are under the care of your physician. The most common kinds are sennosides (Senokot, Ex-Lax, Smooth Move Tea), bisacodyl (Dulcolax, Correctol), cascara sagrada, casanthranol, castor oil and aloe. Bisacodyl is available as a suppository or tablet. The tablet form usually takes six to ten hours to work. Beware: In my practice, I have on a few occasions seen ischemic colitis (a lack of blood to the colon) with bleeding occur after taking bisacodyl tablets. Senna, cascara and casanthranol are converted by the bacteria in the colon to an active substance that stimulates colonic contraction. Castor oil, although not used very often, works differently than the above laxatives. It works in the small intestine, causing an accumulation of fluid and then evacuation. It does its job quickly, often within two hours.

What is a colonic and colotherapy (colon hydrotherapy)? Is it like an enema? Does it work?

A colonic is different from an enema. An enema is usually (or can be) self-administered. A colonic is the irrigation of part of the colon through a device to control the flow of water by a person who has been trained in the technique. Most colotherapists infuse a slow flow of warm water into the colon through a closed, sterile tube, which is followed by gentle gravity drainage. Often this is accompanied by abdominal massage to move the stool toward the lower bowel. The stool and water are collected in a closed system and removed. The process takes about forty-five minutes to an hour. The water can be infused via a gravity system, or it can be infused under pressure. Up to thirty-five gallons of water may be infused by some practitioners. After completing the hydrotherapy, you sit on a toilet and evacuate the rest of the fluid and feces.

It is *extremely* important that the equipment, including the tubing, be sterile! Complications have rarely been reported, although many years ago there were reports of an amoeba infection transmitted by the procedure and the equipment creating a hole in the colon. However, in recent years very few reports of major problems have occurred. If a large volume of water is introduced into the colon and not removed, it is possible that the electrolytes in the body could become abnormal, causing arrhythmias of the heart.

If you undergo colotherapy, it is imperative that the practitioner be experienced and that the equipment be sterile for each client. Colon hydrotherapy is currently not specifically regulated in most states. Colonic irrigation devices are regulated by the FDA. This procedure has been reported to be helpful in patients with chronic constipation, and even in patients with problems getting the stool out of the rectum through the anus, but good studies showing benefit are lacking. Hopefully large placebo-controlled trials will be done in the future to further explore the effectiveness of this technique for constipation and IBS with constipation. I do recommend colotherapy for some women with constipation. Many of these women have had no success with laxatives. The colotherapy helps "wash" the stool out of the bowel. Occasionally after treatment the bowel function seems to improve on its own.

When do you do tests for constipation, and what do they tell you?

Many physicians are unaware of the tests that are possible and helpful for people with constipation. Your dietary history and symptoms are often clues to what may be going on and if any tests should be done. If your diet contains adequate fiber and you still can't go, then other testing may be necessary.

A good rectal exam is important for everyone with constipation. This exam shouldn't be painful, unless you have hemorrhoids or a fissure. Unfortunately, some people *do* experience pain during a rectal exam. Oftentimes a physician roughly whips his or her finger around in the rectum, which at the very least may cause discomfort. If you have been physically abused in the past, a rectal exam may also be uncomfortable.

A good rectal exam for constipation includes testing for the strength and relaxation of the anal sphincter, feeling whether your pelvis can relax when you are pushing the stool out—but not too much—and that you don't have a pouch in the rectum (a rectocoele) that could trap the stool. The physician will also feel for growths. Often the rectal exam or your history can direct further testing. If you have chronic constipation and you feel that the stool comes down but just won't come out, it is necessary to evaluate pelvic floor functioning with a test known as anorectal motility, which measures your ability to evacuate a balloon from the rectum and measures the pressures in the anus at rest, with straining and with evacuation.

Another helpful tool is an X-ray test, called a defogram, in which barium is put through a tube until it fills the rectum and sigmoid colon, just above the rectum. In women, barium paste is also put into the vagina so the relationship of the vagina and colon can be observed. Films are taken while you evacuate the barium.

If you feel like the stool never even makes it down into the rectum, or seems to get "stuck" higher than your rectum, then your doctor may order an X-ray of the abdomen five days after you swallow a sitz-marker capsule (they're easy to swallow), which contains twenty-four radio-opaque markers. This may allow an assessment of the overall movement of the colon and sometimes the small bowel. This test lets your doctor know whether there is a problem in motility in part or all of your colon. After five days a normal person should have no more than one or two markers left in her bowel. Anyone over fifty should have or have recently had a colonoscopy. Anyone with acute constipation or any of the warning signs discussed above, no matter what their age, should likely have a colonoscopy and possibly an abdominal/pelvic CT scan.

Constipation hurts—and allows me to get plenty of bathroom reading done—but can it be serious?

Sometimes constipation can lead to complications. A common one is hemorrhoids, which are caused by straining to have a bowel movement, or anal fissures—tears in the skin around the anus—which occur when hard stool stretches the sphincter muscle. As a result, rectal bleeding may occur, appearing as bright red streaks in the stool. Warm baths, ice packs and the application of a cream containing an anesthetic and often a steroid (for example, Preparation H, Anusol, Anusol-HC, Analpram) to the affected area help to ease hemorrhoid pain. Treatment for anal fissures may include stretching the sphincter muscle, applying compounded nitroglycerin or nifedipine cream, or surgically removing the tissue or skin in the affected area.

Sometimes straining causes a small amount of intestinal lining to push out from the anal opening. This condition is called rectal prolapse, and it may lead to secretion of mucus from the anus. If rectal prolapse occurs, even fixing the underlying cause (which is important), such as straining and coughing, may not improve the situation. You should consult a colorectal surgeon if the problem persists to see if any further treatment is necessary. Headaches can also be a consequence of severe straining. Usually these will go away shortly after completion of the effort. However, rarely, the headache can be an indication of a more serious condition. Therefore if the headache is severe and/or persists, consult your physician.

Fecal impaction may occur when the stool becomes hard and desiccated, filling up part of the colon and/or rectum. The normal contraction of the colon is unable to push the stool out and the stool becomes "stuck." Fecal impaction occurs most often in children and older adults. An impaction can be softened with mineral oil taken by mouth and by an enema. (Again, mineral oil should not be taken later than with lunch, and after taking it, you should not lie down.)

I find that I have to push between my rectum and vagina or put my fingers in my vagina to have a bowel movement. What is going on?

You may have an outpouching of the rectum that pushes forward into the vaginal wall, which is called a rectocoele. This often occurs after a person has had constipation with a very large stool or strains to get the stool out. When a woman has this condition, during evacuation it is often helpful to place her fingers just inside her vagina and gently push toward the rectum and spine so that the stool gets lined up with the anal opening, easing the passage of the stool out of the rectum.

The stool comes down to my anus but then won't come out. I was told I should have biofeedback. What is that?

Biofeedback will train you to improve the coordination of your pelvic floor muscles during attempted defecation, which may help you have a bowel movement properly. It often involves massage of the pelvic muscles by a therapist as well as "retraining." "Obstructed defecation," when you feel that the stool comes down and won't come out, is due to an inappropriate or inadequate contraction of the muscles that allow the stool to leave the rectum or to an inappropriate contraction, instead of relaxation, of the external anal sphincter, over which you have control. In one type of training, biofeedback sensors are inserted into the anus or taped to the skin in the area around the anus to detect the electrical signal for defecation. By watching the recordings, you'll learn how to relax the pelvic floor muscles during defecation and then learn to keep those muscles relaxed while increasing the pressure at which you bear down in your abdomen. Through this technique and/or with a pressure catheter in your anus, you will learn to relax the anal muscles.

What are some ways that I can relieve constipation at home, without having to run to my GI specialist right away?

- Eat a well-balanced, high-fiber diet that includes beans, bran, whole grains, fresh fruits and vegetables. If this makes you worse or gives you more gas or discomfort, cut back on some of the fiber.
- Drink plenty of liquids.
- Exercise regularly.
- Set aside time after breakfast or dinner for undisturbed visits to the toilet. However, don't sit on the toilet for long periods of time if you can't go within a short period.
- Do not ignore or resist the urge to have a bowel movement. When you have an urge, this is most likely the right time to produce a good stool.
- Try raising your feet up four to six inches by resting them on a couple of books or a stool. This helps straighten out the angle of the bowel and makes it easier for the stool to come out.
- Whenever a significant or prolonged change in bowel habits occurs, check with your doctor.

I find that if I have a cup of coffee in the morning, I'd better be near a bathroom! Is this common?

Not everyone has an urge to have a bowel movement after drinking coffee, but a significant number of people do (almost two-thirds of women in one study). Coffee stimulates the anal sphincter and increases the motility of the lower colon, thereby resulting in a desire to have a bowel movement. Therefore if coffee affects you, and you drink coffee in the morning, be sure to leave extra time to use the toilet before rushing off to work. But beware: coffee can dehydrate you and can make it more difficult for you to have a bowel movement, too. Many doctors suggest that you stay away from coffee if you are constipated. This clearly depends upon the effect it has on you.

Can constipation ever be a sign of irritable bowel syndrome?

Yes. Diarrhea isn't the only symptom. IBS means any change in bowel habits with abdominal pain, not just the runs. In fact, IBS with constipation has about the same frequency as IBS with diarrhea. Many women who suffer from long-standing, frequent constipation actually have IBS.

WHAT YOU NEED TO KNOW:

1. Normal bowel movements are different for everyone.

2. Constipation is defined as fewer than three bowel movements per week, a decrease in the frequency of or amount of your normal stools, less than one hundred grams (about a quarter pound) of feces, or difficulty getting the stool out.

3. Fiber may be helpful in some people with constipation.

4. Tests for evaluation of your constipation are available and could be helpful in treatment.

5. Over-the-counter treatments for constipation are readily available, but you should consult your doctor about their long-term use. Your doctor may have other medications or suggestions for treatment if your constipation persists.

6. If you have any bleeding, weight loss, anemia, poor appetite, weakness or sudden constipation, see your doctor.

Chapter 6

Stinky Burps: Heartburn and Halitosis

*"I would like to find a stew that will give me heartburn
immediately, instead of at three o'clock in the morning."*
—John Barrymore

You're lying in bed a couple hours after a desperate late-night stop at Taco Bell when suddenly a burning sensation creeps up your throat. You swallow, and the fiery sensation only gets worse. You finally bolt for the refrigerator and gulp a glass of milk to put out the flames. Relief, at last. This hot demon is acid reflux—formally known as gastroesophageal reflux disease—a burning, acidic splashing in the throat usually brought on by acidic treats like salsa or red wine. Sometimes we can just pop a Rolaids and feel better. But for some of us, the flames never stop burning.

GERD is prevalent in women. Why? Well, we women are always on the go, often taking care of everyone else but ourselves. We eat on the fly, scarfing down whatever's in reach, and give ourselves indigestion. Or else we swill coffee to get us through the day—another indigestion culprit—and end up eating late at night and then lying down in bed, causing the food to come back up. Also, many of the medications we take cause problems with the esophagus and stomach. Medications for osteoporosis—Fosamax, for example—are particularly prone to cause inflammation of the esophagus and a burning in the middle of your chest. Even simple over-the-counter medications like Advil can cause symptoms.

Not surprisingly, we spend plenty of money to squelch the flames. Americans spend more than ten billion dollars per year on proton pump inhibitors (drugs like Nexium and Prilosec); they make up more than 50 percent of total prescriptions. Between 10 to 20 percent of people in the Western world suffer from reflux. So while you might loathe the burning, you're definitely not alone in feeling it.

Are heartburn and GERD the same thing?

Not always! In most cases, heartburn is caused by acid refluxing up into the esophagus and causing a burning feeling, but it could also be caused by an ingested food or medication causing a burning on its own without acid participating. GERD is the process of *any* fluid or food coming up the esophagus, not just acid. It could be bile, recently eaten food or drink.

What exactly is that burning I feel whenever I get indigestion? I'm picturing a pool of evil-looking stomach acid creeping up my throat like a bubbling cauldron.

The burning behind the breastbone and up into the throat is often hydrochloric acid, which is normally made in the stomach to help digest food. More acid is made after a meal because acid is used to help us digest food. The acid helps protect us from infectious organisms, swallowed with food and water, which gain entry into our bodies through our intestines. It also helps start the digestion process by creating an environment where pepsin in the stomach starts breaking down our food and stimulates the release of other important substances that continue the food digestion in the intestines.

Some people have an especially sensitive esophagus, meaning that they feel the acid more readily, even though they might not have any inflammation that the acid can cause. Basically, everyone has some degree of acid reflux; not everyone feels the burn. Some people feel burning but don't get inflammation. Some people get both burning *and* inflammation. It's also possible for fluid or food to come up, not necessarily acid. Only 5 to 10 percent of acid reflux actually causes heartburn.

What causes heartburn?

Heartburn happens when acid flows up into the esophagus and diffuses into the wall of the esophagus, causing pain. This occurs when the lower esophageal sphincter (LES)—the natural valve that keeps stomach acid in the stomach and out of the esophagus, located where the esophagus and stomach meet—relaxes or malfunctions. When functioning normally, the LES opens like a one-way door, allowing food into the stomach but not out the same way. But sometimes the LES sleeps on the job. It relaxes and allows stomach juices to flow upward into the esophagus.

The LES works with gravity's help, which is why it's so common to get heartburn at night, when you're lying down. This relaxation exposes the esophagus to the harsh acid from the stomach. A small amount of acid coming up into the esophagus at times is normal. That's why a majority of people feel heartburn at some time. When acid reflux happens, the esophagus usually clears the acid again, and down it goes back into the stomach, but not so for people with GERD. The esophagus may not clear the acid or food contents very well, and they may stay for a long time in the esophagus, causing more symptoms or damage.

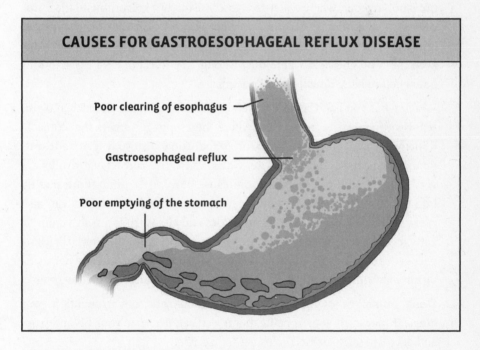

CAUSES FOR GASTROESOPHAGEAL REFLUX DISEASE

Poor clearing of esophagus

Gastroesophageal reflux

Poor emptying of the stomach

Figure 6-1. Several mechanisms contribute to gastroesophageal reflux disease (GERD). The first is the backward motion of food and/or acid out of the stomach into the esophagus through the high-pressure area where the two join—the LES. The second occurs if the esophagus does not clear the acid and food contents back down into the stomach; the material can cause damage to the esophagus or get into the mouth or down into the lungs. The third occurs if the food and liquid are not cleared well from the stomach and just sit there, allowing for an increased risk of reflux back into the esophagus.

Esophagitis is the inflammation (with or without ulcerations) of the esophagus caused by gastric acid and food contents. It can also be caused by pills and infection.

Are there any natural ways to prevent or treat my heartburn?

The most important treatment for heartburn is lifestyle modification. Take a multipronged approach:

1. **Sleep sensibly.** The head of your bed should be elevated about six to eight inches.

2. **Time your eating.** The timing and volume of dinner can also contribute to reflux or heartburn. If you have a lot of food just sitting in your stomach and not emptying out, where will it go? Nowhere but up! I recommend the European style of dining (minus the late night dinner)—eat your biggest meal at lunch, and don't eat too much at dinner or before bed. After all, gravity helps keep the food and acid where they belong. Try to eat at least three to four hours before lying down, if at all possible.

3. **Watch what you eat.** Caffeine increases acid output in the stomach and also reduces the pressure in the LES, where the esophagus enters the stomach. Caffeine doesn't lurk only in tea or coffee. You'll also find it in soda, migraine medication and chocolate. Colas and other carbonated drinks are very acidic, and the carbonation can cause burping, sending more acid up into the esophagus. Garlic, onions, mint, fatty foods and alcohol can also cause lower pressure in the LES; tomatoes and citrus fruits are also culprits. Fatty foods and high-fiber foods are slow to empty from the stomach. Many people drink milk, thinking that this might neutralize the acid and help. Unfortunately, milk may stimulate acid and can potentially make things worse.

4. **Don't smoke!** Smoking is bad for your esophagus, not to mention your lungs. It may curb your appetite, but it also can decrease your LES pressure and decrease saliva, which helps neutralize acid.

5. **Lose weight.** Obesity increases your risk for not only heartburn and reflux but also ulcers in the esophagus and cancer of the esophagus. Your body mass index is directly related to the number of reflux events you have: the heavier you are, the more acid reflux events you'll have. This may be due to increased pressure on the stomach or more episodes of relaxation of the LES.

6. **Avoid tight clothing.** Luckily for all of us, corsets are no longer in style. However, body shapers, like Spanx, are in. These can cause a lot of pressure on the stomach, as anyone who's ever wriggled into one can attest. By increasing pressure on the stomach, food and acid may reflux up through a weak LES.

7. **Analyze your meds.** Medications often cause burning or inflammation in the esophagus. Prime offenders are potassium and tetracycline. Nonsteroidal anti-inflammatory drugs (NSAIDs) like ibuprofen, naproxen and aspirin can cause ulcerations in the stomach, duodenum or small intestine and can also cause inflammation in the esophagus. If you must take medication, note that coated aspirin might help; taking your medication with misoprostil (used for ulcer prevention), with a proton pump inhibitor (PPI) or with food is also helpful.

   Many women take bisphosphonates like risedronate (Actonel) and alendronate (Fosamax) for osteoporosis; these can cause inflammation in the esophagus and heartburn and pain. I tell my patients to avoid lying down after taking them; some doctors also administer a PPI to take alongside these medications. Be sure to take any medication with plenty of water.

How do medications work to cure my heartburn?

There are three types of medications to treat heartburn. **Antacids** remedy the occasional heartburn, neutralizing the acid made in the stomach. They work quickly and can give fast relief, but they don't last long and may need to be taken frequently. Antacids contain combinations of aluminum, magnesium, sodium bicarbonate and calcium. Liquid antacids are more effective than tablets, because the medication is more evenly dispersed over the lining of the stomach (though they're not as easy to carry on the go). You should know that an antacid high in magnesium may cause loose stools or even diarrhea; antacids high in calcium or aluminum, meanwhile, are more likely to cause constipation.

The second type of therapy for heartburn is **H2 (histamine type 2) blockers**. Acid is made by the parietal cells in the stomach. Histamine is a primary stimulus to this cell to secrete acid. If the receptor for the histamine is blocked on the cell, then less acid will be released. Some acid will still be made in the stomach with this type of medication, though, so H2 blockers might not help everyone; it's been shown to help 10–24 percent of people who don't get better naturally or through use of a placebo. Also, H2 blockers don't help heal severe inflammation of the esophagus very well. However, for mild symptoms, they can be worthwhile to decrease acid and symptoms. If they don't work after six weeks, then you should change to the third type of therapy, **proton pump inhibitors (PPIs)**, which I discuss below.

Many of the H2 medications are available over the counter but may be cheaper if you get them via prescription. The dose of the medication depends on the severity of your symptoms. Whereas only the weaker formulations were initially available over the counter, now the more potent ones are also available. It's important to realize that H2 blockers can interfere with the function of many other medications, affecting their potency. It's crucial to check with your doctor or pharmacist to inquire about possible interactions.

PPIs are the most potent form of therapy. These drugs directly block the secretion of the acid in the parietal cell when this cell is turned on to secrete acid. Think of the cell as a faucet—when it's on, it leaks acid; when it's off, it doesn't. PPIs bind to the switch, turning the cell to the "off" position, thereby stopping (if completely effective) acid production.

The acid pump flows at its strongest after you eat. Therefore, you should take your PPI about a half hour to one hour before you eat so it has enough time to be absorbed by your body and, by way of circulation, to attach to the parietal cell. The most effective time to take the medication is before breakfast. Usually one dose of medication is enough to stop the heartburn and heal any ulcerations. Most people take one PPI in the morning and the effects last all day, but in about 10–15 percent of people a second dose is necessary before dinner. It takes about five days for the acid secretion to stop. So be forewarned: You might not get the same relief if you take PPIs on an as-needed basis. Give them enough time to work.

As for which PPI is best, rabeprazole may last a little longer than the other conventional PPIs, but this doesn't usually matter for most people. I recommend that a person first try taking the cheapest PPI and see if it helps. Which PPI is the cheapest for you will depend on your health insurance plan. A prescription of generic omeprazole, for instance, is usually cheaper than Prilosec, which is available over the counter. In some individuals PPIs stay around longer and have a longer-lasting effect. In other individuals PPIs are broken down (metabolized) more quickly and aren't as effective. Another reason to switch PPIs may be a side effect. The most common side effects of PPIs are abdominal pain, diarrhea, headache and possibly constipation. If you have these side effects from one PPI, you may be able to tolerate another without the same side effects.

Will I have to take medicines for my whole life in order to prevent heartburn?

Hopefully not. Unfortunately, acid reflux often comes back and may require more medications. The best thing to do is to try to avoid the foods that give you acid reflux and maintain the other lifestyle changes I mentioned above. I like to try to reduce the dose of the medications for GERD after a patient is symptom-free for two to four weeks. I first try to reduce the dose of the medication; then if someone is on a PPI, I try to change over to an H2 blocker.

Unfortunately, if you've gotten ulcerations in your esophagus from the acid, this approach may not work. If you've had long-term heartburn, get an endoscopy to check for ulcers or a change in the lining of the esophagus, a condition called Barrett's esophagus, which can, in a small number of people, progress to esophageal cancer. But just because you suffer from heartburn doesn't mean you're destined for an ulcer, which happens when the lining of the esophagus becomes eroded. There's more than just acid involved; proteins and enzymes also combine to form ulcers, helped along by the acid but not caused by it. If you've had ulcers or Barrett's esophagus, I recommend you stay on a low-dose PPI.

However, if you haven't had either condition and have had only mild symptoms, you might get away with stopping your medicines and using them intermittently when and if your symptoms return. You have about an equal chance of remaining symptom-free or needing to take another course of medication.

Lately I've seen frightening headlines cautioning against common remedies like Prevacid and Prilosec. How come?

When PPIs were first approved by the FDA, it was only for three months of treatment. There was a fear that PPIs might confer an increased risk for cancer in the stomach. To date, fortunately, that hasn't happened. However, other possible problems have been reported with long-term use of PPIs. Decreasing or eliminating the stomach acid causes a problem with the absorption of certain nutrients that require acid in order to be optimally absorbed into the body, like calcium, magnesium and vitamin B_{12}. A modest increased risk (1.45/1.25 times) of spine fracture and wrist fracture (but not hip fracture) and slightly increased risk of osteoporosis in women over fifty on PPIs has been reported, and the likelihood of this increases for women on higher doses of PPIs. We don't know if this is due to calcium absorption alone or also an effect on the bones directly. In addition, we don't know if taking a higher dose of calcium or a different calcium supplement with adequate vitamin D can reverse the slight increase in osteoporosis. Calcium citrate, which is better absorbed than calcium carbonate if there isn't acid in the stomach, would be a better supplement choice in people on a PPI. Both supplements should be taken with food to enhance their absorption into the body. At this time, there is no recommendation for how frequently a bone density test should be done if you are taking a long-term PPI.

If you're deficient in compounds such as zinc, vitamin B_{12}, vitamin C or vitamin B_6, you should be taking vitamins to correct the problem. Vitamin B_{12} levels should be monitored by a blood test if you're on a long-term PPI. You may need vitamin B_{12} supplementation, which may be given orally, by nasal spray or by a shot.

Other potential problems, which can even exist during short-term use of PPIs, include a higher risk of traveler's diarrhea or food poisoning. If you don't have the acid to kill off some of those ingested bad "bugs," you're more likely to get sick. Increasing acid suppression is associated with an increased risk of developing *Clostridium difficile* infection (usually from antibiotic treatment). Furthermore, if you develop an infection with *C. difficile* you may have more difficulty getting rid of the infection, with it coming back after the usual treatment. PPIs also might make you more susceptible to developing pneumonia both in the hospital and at home within the first month of starting the medication.

A recent report suggests that a commonly taken medication, clopidogrel (plavix), for cardiac disease might be less effective if the person is also taking omeprazole. This finding has been contradicted by other reports. More information is needed to determine the accuracy of this interaction. Given what we know at this time, if your doctor thinks that you need a PPI, you should continue taking it. PPIs can also affect the metabolism of certain other drugs, such as warfarin (Coumadin), digoxin, phenytoin (Dilantin), theophylline (used for asthma), diazepam (Valium) and carbamazepine (Tegretol). Make sure that you check with your doctor or pharmacist about the possible interactions of your medications. Omeprazole (Prilosec) is the most likely PPI to interact.

I wake up in the middle of the night, coughing, and sometimes my food from dinner (and even lunch!) comes back up. It can be really scary. Sometimes I have to sit straight up in bed and cough out the fluid and food in order to catch my breath. Why does this happen, and what can I do about it?

When you lie down at night, you don't have gravity to help prevent the food or liquid in your stomach from coming back up the esophagus. If your LES is weak, then reflux up the esophagus and even into your mouth can be the result.

Sometimes old food can lurk in the body if you have Zenker's diverticulum, which is an outpouching of the esophagus. This diverticulum is in the upper part of the esophagus, and some food can sneak into it and get trapped, and then come back up when you lie down at night. Surgery can fix it; luckily, this isn't a common problem.

Old food can also come up if your stomach isn't emptying normally. Abnormal emptying occurs in up to 55 percent of people with type 1 diabetes and in about 30 percent of people with type 2 diabetes. Poor stomach emptying can also happen after an intestinal infection. Although this is usually short-lived, it can persist for a long period of time in some, especially women. Up to 30 percent of people with IBS may have some delay in gastric emptying. Medications or previous surgery also can be culprits. Common medications that cause this problem are anticholinergic agents (drugs that block the action of the neurotransmitter acetylcholine, like dicyclomine, atropine and hyoscyamine), narcotics, tricyclic antidepressants and calcium channel blockers (agents that prevent calcium from entering the heart and walls of cells of the blood vessels, used for treatment of high blood pressure, chest pain and irregular heart). Alcohol and nicotine also delay stomach emptying.

If you experience reflux into your mouth or coughing at night, see a doctor to rule out an underlying cause. An upper endoscopy can make sure that there isn't a cancer or scarring preventing the food from emptying out of the stomach. Inflammation from an ulcer or gastritis can also cause the problem. If an upper endoscopy shows scarring in the opening leading out of the stomach, this may be able to be dilated with a balloon or even injected with Botox. These procedures have provided short-term relief in some people.

If the problem is motility, therapeutic options are limited. Stopping medications that might affect motility is important. Then the diet should be modified: eat smaller meals more frequently, increase liquids and avoid fatty or fibrous foods, which empty from the stomach slowly.

My asthma has been horrible this year. My doctor says that acid reflux may be aggravating it. Does that really happen? The two don't seem connected, and I don't feel symptoms of reflux.

Reflux of acid or other contents from the stomach that get into your lungs or larynx can likely cause worsening of asthma. If you have symptoms of acid reflux, you should take a PPI medication and see if your asthma feels better. However, a recent study in the *New England Journal of Medicine* of men and women, but mostly women, who had uncontrolled asthma and few symptoms of acid reflux reports that they were not made better by taking Nexium daily. In that study about 40 percent (two out of five) of the asthmatics on the PPI and on the placebo had demonstrated acid reflux by pH probe testing. So, if you do *not* have symptoms of acid reflux, I would first recommend that you avoid eating large meals at night and eating shortly before bedtime, since there is no proof that taking a PPI will improve your asthma. On the other hand, if you *do* have reflux symptoms, then a PPI might be beneficial.

When should I get tests to look at my esophagus and stomach?

If you have trouble swallowing (food gets stuck going down or goes down slowly), still have symptoms after four weeks of a PPI treatment (used twice a day if once a day failed), are anemic, have severe pain, have chronic nausea, feel like food won't leave your stomach or are losing weight, get an upper endoscopy. An endoscope, a tube with a camera, will be inserted into your mouth and advanced into your esophagus, stomach and duodenum and should give a clearer picture of what's happening.

I know esophageal cancer is extremely serious, and I'm worried that I might get it, thanks to my reflux. How do I know if I have cancer?

If you have persistent, unresponsive symptoms of heartburn or reflux, or trouble swallowing, with or without weight loss, your doctor will do an X-ray study or a gastroenterologist will do an upper endoscopy. A cancer of the esophagus and stomach can be seen on those tests. Pieces of tissue from the tumor or an ulceration that is seen will be examined under the microscope for tumor cells to see what kind they are in order to determine the correct treatment.

Is there anything I can do to prevent a cancer of the esophagus?

Stop smoking and drinking alcohol, and maintain a healthy weight. Smoking, excessive alcohol and obesity are associated with an increased risk of esophageal cancer. If you have had persistent or recurrent symptoms of heartburn, then you should have a gastroenterologist do an upper endoscopy to examine your esophagus. He or she will also look for Barrett's esophagus. This is a condition where the lining cells of the esophagus, when damaged by acid, change into intestinal-lining cells. This condition is more common in Caucasian men but can occur in women. It causes an increased risk for developing cancer, especially if it is extensive. If you have this condition, you should take a PPI to try to keep the acid suppressed, and you should have repeat endoscopies, done at regular intervals.

Can I have surgery to cure GERD?

Surgery is available for the treatment of GERD and most is done via a laparoscope these days. However, long-term studies suggest that surgery might not be better than medication for the treatment of acid reflux. Surgery may be the right step if you've had:

- Ulcerations of the esophagus that keep coming back or don't heal well with adequate medication.
- Persistent heartburn on maximum PPI medication with documentation by testing that you make acid and your heartburn occurs when you have acid or other fluid refluxing.
- Regurgitation with fluid coming up into your mouth, with coughing, trouble sleeping or aspiration pneumonia.

- A problem with asthma due to regurgitation of fluid or acid.
- Barrett's esophagus with abnormal cells. It is important to do a pH test to make sure that the heartburn that you may be feeling actually corresponds to the presence of *acid*. None of the long-term studies comparing medication and surgery to date have shown a difference in the development or progression of Barrett's esophagus or the development of cancer. And don't expect to be medication-free at ten years. In one study only 35 percent (mostly male study participants) were medication-free.

I still have heartburn, even though I take my Prilosec twice a day. What can I do about it?

Make sure you are taking your medication before meals and that you regularly take your medication. Also, watch what you eat—that means no supersized coffees to propel you through the day. If your medication still isn't working, your esophagus should be examined via an endoscope to see if you have ulcerations or inflammation. If you do have inflammation, then increasing the PPI, changing the PPI to a different one or possibly taking an H2 blocker at night in addition to the PPI may help. Even if you have a normal endoscopy, it is worth trying an increased dose (40

twice a day) or changing to another PPI or adding an H2 blocker at night.

If after four weeks you're still not better, then it's time for a pH test of your esophagus. This is done in two ways. The first test is done with a small monitor about a tenth of an inch wide that is passed through the nose and into the stomach. It's connected to a small, easily portable box. It can measure the pH of the upper and lower esophagus, which indicates the levels of acid and base. Many of the probes can determine whether there is fluid coming up the esophagus (measured by impedance [changes in electrical resistance detected by the probe]). The probe is usually left in place for twenty-four hours so that it can record what's happening in your esophagus during normal activities—eating, exercising—during stressful events and during the night. During the monitoring, you indicate when you experience heartburn, reflux or a cough, and you keep track of what you're doing throughout the day. The monitor is generally well tolerated. The second test involves the use of a wireless capsule. This capsule is attached to the esophagus wall during an endoscopy. Monitoring of the pH in this location in the esophagus is usually done for forty-eight hours. The capsule transmits the values by radio signals to a receiver worn on your belt. Eventually the capsule falls off the esophageal wall. The presence of the capsule is generally well-tolerated, although as many as 25 percent of people will complain of increased pain due to its presence. The test is done when you are either on or off of your medications.

Testing for the presence of the bacterium *Helicobacter pylori,* which is linked to inflammation of the stomach lining, is not necessarily needed if you simply have heartburn, because its treatment in this case is controversial. Although it was suggested in the past that treating *H. pylori* worsens GERD symptoms and complications, the current evidence does not support this view. At this time, there's no evidence to suggest you should be treated for *H. pylori* in order to treat your GERD symptoms.

Can my body become resistant to my PPI?

It's possible. After being on a PPI for two months, acid hypersecretion can occur, which means extra acid can be released in the stomach. This can persist for over two months. A recent study suggests that PPIs may even cause or aggravate heartburn after use. After normal people without GI symptoms finished a two-month course of esomeprazole (Nexium), 40 mg per day, nearly half (44 percent) had clinically significant heartburn, acid reflux or dyspepsia, compared to 15 percent of people on an inactive compound (placebo). These symptoms persisted for at least four weeks. Therefore, it may be hard to get off a PPI; I recommend people go off gradually in conjunction with an H2 blocker.

One of my most confounding GERD cases involved Heather, a high school athlete suffering from acid reflux with regurgitation, lactose intolerance and irritable bowel syndrome with diarrhea, controlled by Imodium. She'd experienced digestion problems since she was about six years old. Because of symptoms in middle school, she had an upper endoscopy, which showed erosions (like canker sores) in her stomach, and she had a colonoscopy, which was normal. A repeat upper endoscopy six months after the first was normal, as were an ultrasound of the abdomen and an upper GI series X-ray. A gastric emptying study was normal for solids, but liquids emptied from her stomach slightly more slowly than normal. Because of reflux symptoms, she was placed on Prilosec (omeprazole) and had been on this medication for several years before seeing me. Toward the end of high school, her reflux symptoms became much worse. When she ate, food seemed to regurgitate up her esophagus, with the sensation that it could even escape through her nose. This usually happened right after she ate, especially after big meals. She switched to a different type of PPI, Protonix (pantoprazole), which helped a bit. To empty her stomach, I suggested that Heather modify her diet. However, she wasn't particularly successful in modifying the way she ate. Even though the emptying study was mildly abnormal, the fact that she regurgitated food that she had eaten up to twenty-four hours before suggested that her stomach emptying at times was very abnormal. She successfully stayed away from most fat in her diet, but she wasn't able to adhere to my recommendation of eating small, frequent meals. She continued to take the Protonix twice a day. She did very well, with minimal symptoms, during her freshman year of college.

However, the following year the reflux problems became a major issue during her lacrosse games. Although she sometimes forgot, she was usually good about taking her medications correctly before meals. Unfortunately, before games she was unable to eat, sometimes only having a bagel or a PowerBar. She'd have to leave games halfway through due to reflux, throwing up dark brown liquid. She treated this with Rolaids or Tums.

Over the next year and a half Heather was inconsistent with taking her medications, taking only the Protonix once or twice a day, maybe a few times per week. She also revealed that she had made herself throw up after almost every meal in order to control belly distension and the feeling of heaviness she experienced after eating. She estimated that about 50 percent of what she ate was vomited up. She was heartburn-free except when vomiting, when she was guaranteed to have heartburn. She would vomit immediately or up to a few hours after eating. Furthermore, she revealed that this had been going on for a few years. Because it embarrassed her, she hadn't wanted to tell me about it in the past.

Heather clearly had an eating disorder. I emphasized to Heather that vomiting was a big contributor to her symptoms and she should try to prevent it from happening. The acid that was regurgitated could cause further damage to the esophagus and could cause the heartburn to last for longer periods of time. I tried to underscore with her how important it was that she take her Protonix regularly. When she took it intermittently, she'd get an acidic feeling in her chest and abdomen. This didn't happen when she stayed on her medication. I also underscored the things she needed to avoid, listed in the box below.

DON'T GET BURNED

To prevent heartburn:

- Avoid caffeine (tea, caffeinated and decaf coffee, chocolate)
- Avoid carbonated beverages
- Avoid alcohol
- Avoid citrus, tomatoes, mint, garlic
- Avoid cigarettes

Over the next year Heather did very well on Protonix, but because she did better, she began taking it infrequently again. Yet again, she had symptoms. At this point she's learned that she absolutely has to take her medications regularly. Just because she feels better, she cannot abandon her treatment. As Heather's case shows, the most common cause of treatment failure is not taking all your medications as you should!

Remember:

1. Communication with your physician is essential so that the proper advice can be given.

2. It's important to take your medications regularly, as directed. If you don't have symptoms and feel that you can stop your medications, you should have a conversation with your doctor so that successful strategies to try to reduce and then, if possible, to stop your medications can be developed.

3. If you have a recurrence of your symptoms, think what you may be doing to aggravate the situation. Are you eating properly? Are you eating foods that cause acid reflux? Are you taking prescription or over-the-counter medications that could be contributing? Are you lying down just after eating? In Heather's case, her symptoms were provoked by her spotty medication intake and induced vomiting.

Sonia, a suburban schoolteacher, had the onset of GERD in her early forties. Because of food getting stuck on the way down into her stomach, she had undergone a barium study of her esophagus and an upper endoscopy within the past seven years. The barium test was normal, and the endoscopy showed a small hiatal hernia and some mild redness where the esophagus and stomach joined, although the tissue in this area was normal under the microscope. She did very well for a year after the endoscopy but then developed severe chest tightness. Her primary care doctor put her on Zantac (ranitidine) and checked out her heart with tests, which reassuringly were normal. Because her doctor thought the chest tightness was likely due to acid reflux, since Sonia's heart appeared to be okay, she put Sonia on Prilosec (omeprazole), which resolved the chest tightness after two weeks. Sonia then stopped the Prilosec and did well, with only rare chest tightness.

However, when I saw her six years later, Sonia now had symptoms. She complained of hoarseness and chest pressure, along with the sensation that food was sticking in her esophagus. She admitted to drinking coffee and eating hot peppers, but she shied away from wine. Her weight was normal, so I knew that wasn't contributing to her problems. She'd started herself back on Zantac to try to control her hoarseness and chest pressure. This wasn't working. I put her back on Prilosec.

After a month of treatment the heaviness in her chest disappeared, but the hoarseness persisted. She continued on the Prilosec for another four months and then switched back to Zantac. Although she had tolerated the Zantac in the past, for some reason it now gave her leg cramps, which stopped when she stopped taking her medication. But without her medications, her heartburn recurred. She ultimately returned to Prilosec.

However, one morning before breakfast she felt like she had lead on her chest, despite normal eating habits and the very infrequent ingestion of small amounts of wine. She did tell me that she routinely went to bed within three hours after eating dinner.

I felt that Sonia was likely having heartburn because she was going to bed shortly after eating a big meal. I recommended that she eat a smaller meal at night, at least four hours before bed, and that she try to avoid wine and coffee. This worked for the next seven months.

Then, one night, Sonia awakened from a sound sleep with severe gas pains under her left breast that shot straight through to her back, between her shoulders. Fearing a heart attack, she went to the emergency room, where a cardiac evaluation came back normal. The previous evening she had had spaghetti sauce made with a lot of garlic, and a gin and tonic. She rarely drank alcohol and could not remember when she had last had hard liquor.

She continued taking Prilosec after this incident, and things were status quo for another year. She avoided dietary pitfalls and was careful not to eat too close to bedtime. Eventually, though, the chest and back pain came back; gradually, the Prilosec's effectiveness had waned. She was started on a different PPI, AcipHex (rabeprazole); it worked well but made her jittery. Finally, I switched her to Nexium (esomeprazole), eventually doubling the dose, which eliminated her symptoms. I also made sure she took enough vitamin D and calcium and had a bone-density test, as these drugs can affect the levels of both substances.

WHAT YOU NEED TO KNOW ABOUT GERD:

1. GERD symptoms can occur over long periods of time. It is possible in some cases to have intermittent therapy to reduce the acid.

2. Chest pain can be a symptom of GERD. However, it is essential that possible problems with your heart, such as angina (chest pain or discomfort that occurs when your heart muscle doesn't get enough oxygen), be ruled out as the cause of the pain. Heart problems can be serious and the symptoms can be the same as those with GERD.

3. Occasionally, you might not tolerate a specific PPI. This does not mean that another one will necessarily give you a similar problem.

4. When taking a PPI, it is very important to have adequate calcium and vitamin D and to have your bone density checked per the recommendations of your doctor.

I heard that stomach ulcers and duodenal ulcers often are caused by an infection. I have had pain and I was told that an X-ray shows a duodenal ulcer. Should I be treated with antibiotics?

If you're shown to have an infection with *Helicobacter pylori,* the type of bacteria that causes ulcers, then yes. If you don't have an infection, I do not recommend treatment with antibiotics.

The two most common causes of duodenal ulcerations are pain relievers known as NSAIDs (including low- and regular-dose aspirin, ibuprofen and naproxen) and *Helicobacter pylori* infection. If you have an ulcer and you're not taking an NSAID, your likelihood of having the *H. pylori* infection is over 70 percent. These two causes of duodenal ulcers are independent, and if both are present, there may be a higher risk of developing an ulcer and bleeding. *H. pylori* takes up residence in the stomach. Thirty to 40 percent of people in the United States are infected with *H. pylori.* If you have a duodenal ulcer and there is a high likelihood that you may be infected with this bacteria, a blood test for the antibodies to *H. pylori* is the cheapest and most comfortable way to determine whether you should be treated. If you had an ulcer in the past, the antibodies may stay positive and will not indicate whether the infection is acute. However, if you have a stomach ulcer, have never been treated for *H. pylori* before and your antibodies are positive, you should be treated. If you are taking NSAIDs, the treatment of the *H. pylori* may reduce your risk for bleeding. The chances of an ulcer-free period at one year are 98 percent if you wipe out the bacteria, compared to 65 percent if the bacteria are still there. Treatment with antibiotics is also better for ulcer healing compared with medication that simply decreases acid.

You have several treatment options. In the United States the recommendation is therapy with (1) a PPI usually twice a day, clarithromycin (related to erythromycin) twice a day and amoxicillin (related to penicillin) or metronidazole (Flagyl) twice a day for fourteen days; or (2) a PPI or H$_2$RA twice a day, bismuth four times per day, metronidazole four times per day and tetracycline four times per day for ten to fourteen days. Only one PPI, AcipHex, used in combination with clarithromycin and amoxicillin is approved for seven-day use. In the United States the recommendation is fourteen days of therapy. These medications are often prescribed individually or as a package containing the three together. Eradication of the *H. pylori* infection is about 70–85 percent.

Any medication can potentially cause an allergic reaction with rash or hives, or, more seriously, swelling or shortness of breath. If this happens, you should immediately stop your medications and call your doctor.

SIDE EFFECTS OF MEDICATIONS USED TO TREAT ULCERS

PPI—headache, abdominal pain, diarrhea

Clarithromycin—nausea, diarrhea, abdominal pain, altered taste

Metronidazole—metallic taste, nausea, abdominal pain, severe reaction when taken with alcohol

Amoxicillin—GI upset with diarrhea or nausea, headache

Tetracycline—sensitivity to the sun with a rash, nausea, diarrhea

---

If you develop an ulcer while on NSAIDs, you should try to stop the medication if at all possible and undergo treatment for your ulcer.

Does GERD cause bad breath?

Oh, yes! This was recently described in a study in Israel of people being treated at a gastrointestinal clinic. Symptoms that correlated with bad breath, from the strongest to the least strong association, were: sour taste, regurgitation, heartburn, rumbling in the stomach and difficulty swallowing. How exactly GERD causes bad breath is unclear. Recently, it was shown that there is a specific bacterial flora (microbiome) in the lower part of the esophagus that can change if there is inflammation (esophagitis) in the esophagus. In a small study the bacteria occupying the distal (lower) esophagus in those people with esophagitis had more gram-negative anaerobes/microaerophiles. As certain anaerobes (bacteria that don't need oxygen to grow) are the cause of bad breath, it's possible that a change in the bacteria of the esophagus could be responsible for the bad breath in those with GERD.

About 10 to 30 percent of people have regular bad breath.

My teenage patient Theresa suffered from a long-standing problem with bad breath. Her woes started at the beginning of high school. Vigorously brushing her teeth and tongue did nothing. She saw her primary physician, who thought that she might have an inflammation of her sinuses and treated her with Sudafed, followed by an antibiotic, without improvement. She then tried hydrogen peroxide mouth rinses, recommended by a dentist. This failed, too. Theresa's orthodontist thought that her braces were the malodorous source. No luck. Sometimes people with tonsil problems have bad breath, but Theresa's tonsils had been removed years ago. Doctors were stumped.

The bad breath very much affected Theresa socially; she tried to avoid standing too close to friends or teachers. Finally, she was sent to see a pediatric gastroenterologist, who put her on Prilosec. At last she improved a little bit. At the same time Theresa was having difficulty going to the bathroom; perhaps increasing her elimination of stool might be the answer. Theresa was started on MiraLAX (a laxative that draws fluid into the bowel). At the same time, she tested positive for bacterial overgrowth in the small intestine or GI tract. The overgrowth was treated with metronidazole, which remedied the bad breath within a couple of days. Two weeks later, though, the odor returned. This time, a combination of metronidazole and the probiotic lactobacillus was tried. Theresa enjoyed a blissful, halitosis-free five weeks, but then, sure enough, the bad breath came back. Finally a different antibiotic, Augmentin, was used and this also didn't work, although a third antibiotic, Cipro, helped a bit. She remained on her probiotics and MiraLAX. To rule out any serious disease, Theresa underwent a colonoscopy, a capsule endoscopy, an upper endoscopy and food allergy testing, all of which were normal. Finally, as a last resort, Theresa was placed on a specific carbohydrate diet (see Chapter 9). This diet limits the ingestion of grains, processed meats, foods with added starch and milk products. These foods contain indigestible compounds that the bacteria use as their own food source, and in the process of ingesting these compounds, the bacteria release smelly compounds.

Since then, Theresa has been symptom free. Her mother questioned whether treatment during her labor with Theresa might have contributed to her condition. Her mother noted that when she was in labor with Theresa, she had an infection. After Theresa's birth both she and Theresa were given antibiotics for four to five days. While there are no studies to confirm it, it's possible that taking antibiotics near and after birth affected Theresa's gut bacteria so that she subsequently got bad breath. It's also possible that Theresa's constipation caused more or different bacteria to take up residence in her GI tract, with the result of an increased formation of the organic substances causing bad breath.

What causes bad breath?

About 80 to 90 percent of bad breath is caused by a source in the mouth. This can be a tongue coating, a disease of the gums, diseases of the teeth (cavities, root disease), the presence of braces with trapped bacteria, food residue, unclean dentures or decreased salivary flow that affects the bacteria. Food can be a source. Garlic and onions are common culprits, of course. We've all tried to hold our breath when standing too close to a garlic lover. "Lobster mouth," after eating lobster, and a fishy smell coming from the mouth after taking fish oil capsules are also common. Anything that decreases saliva also increases bad breath, particularly antidepressant and antipsychotic medications.

What can I do to get rid of my bad breath?

First and foremost, brush the back of your tongue and the roof of your mouth, as well as your teeth. The back of the tongue is the most common location for bad-breath bacteria. Avoid foods that cause bad breath—or at least eat them at your own risk! Mouth rinses can also be helpful. And, of course, please visit your dentist. Your dentist will do a good oral exam and decide if areas of the oral cavity need attention. You may need to decrease medications that cause less saliva production.

If your mouth holds no answers, it's time to look elsewhere. Severe liver disease and kidney disease can be sources of bad breath but are usually associated with other symptoms. If other symptoms accompany the bad breath, you should see your doctor. If you have reflux symptoms, then you could try an over-the-counter medication such as an H2RA or a PPI for two weeks to see if that eliminates your reflux and halitosis.

If these approaches to eliminating bad breath fail, figure out if you're actually experiencing a bad taste in your mouth—not bad breath. Then, there are two possible approaches: change the bacteria in the gut or change the food on which the bacteria thrive. At the time of this writing, there are no studies to verify the effectiveness of these approaches, though they have shown to be effective for some patients. To change the bacteria in your gut, probiotics could be helpful, though it's a trial-and-error process.

A more radical way to change the bacteria in your gut might be a short course of antibiotics. However, which antibiotic to use is uncertain, and to be honest, the effectiveness of this approach hasn't been verified. Perhaps those antibiotics that treat bacterial overgrowth, tetracycline or rifaximin, might be useful and certainly are worth a try when you are desperate and nothing works. Diet modification is another option. This was successful for Theresa, who got better on a specific carbohydrate diet. See Chapter 9 for more details.

WHAT YOU NEED TO KNOW:

1. Acid reflux occasionally occurs in all of us, but if it is a frequent occurrence, you may need treatment.

2. Lifestyle changes may help some people: Avoid or decrease your intake of coffee, other caffeine, alcohol, mint, onions, garlic and fatty foods, and refrain from eating shortly before bed.

3. Antacids give fast relief; histamine type 2 (H2) blocker medications provide relief for some people.

4. Proton pump inhibitors (PPIs) can provide long-lasting relief but may be hard to stop. They are needed to treat ulcerations in the esophagus and elsewhere in the stomach and duodenum.

5. Ulcerations caused by *Helicobacter pylori* respond well to antibiotics and proton pump inhibitors.

6. Delayed emptying of contents from the stomach can cause a full feeling in the upper belly as well as contribute to reflux.

7. Bad breath is usually caused by a dental or mouth problem, but it may be caused by acid reflux and by the foods you eat.

Chapter 7

When It's Really Bad: Time to Get Help

*"When the pain is great enough,
we will let anyone be doctor."*
—Mignon McLaughlin,
The Neurotic's Notebook, 1960.

How do you know when it's time to seek a second opinion for more severe issues? This chapter focuses on some of the especially troubling GI problems I see in my practice, such as inflammatory bowel disease and colorectal cancer. I'll also discuss some common problems that are often improperly diagnosed, such as gallbladder disease and diverticulitis.

Often the first clue that something is amiss is belly pain. Belly pain doesn't always mean GERD or irritable bowel syndrome. In fact, IBS is a diagnosis of exclusion—basically, other problems have to be ruled out. The location and character of the pain are often clues; however, it's important to remember that if you have severe pain, you must seek attention right away. Please, don't be your own doctor. Why? Sudden or progressively increasing pain over a few hours or days could be appendicitis, gallstones, inflammation of the gallbladder (cholecystitis), ulcer disease, inflammatory bowel disease or diverticulitis, obstruction of the bowel, kidney stones or ovarian cysts. You can't be your own diagnostician. Most of these conditions need immediate attention and in some cases may even require surgery.

ABDOMINAL PAIN LOCATIONS: WHAT MIGHT BE THE CAUSE?

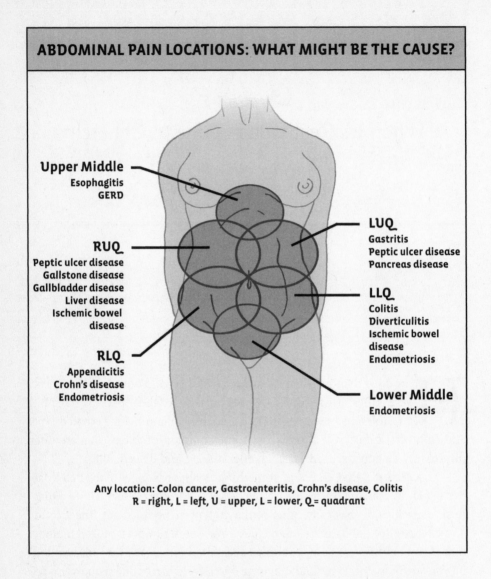

Upper Middle
Esophagitis
GERD

LUQ
Gastritis
Peptic ulcer disease
Pancreas disease

RUQ
Peptic ulcer disease
Gallstone disease
Gallbladder disease
Liver disease
Ischemic bowel
disease

LLQ
Colitis
Diverticulitis
Ischemic bowel
disease
Endometriosis

RLQ
Appendicitis
Crohn's disease
Endometriosis

Lower Middle
Endometriosis

Any location: Colon cancer, Gastroenteritis, Crohn's disease, Colitis
R = right, L = left, U = upper, L = lower, Q = quadrant

Figure 7-1. This drawing shows the locations in the abdomen where pain may occur and lists the most likely digestive tract causes of pain for each location. The order of the letters is (1) *R* or *L* for right or left; (2) *U* or *L* for upper or lower; (3) *Q* for quadrant. The middle quadrants of the left and right sides are not labeled but could be considered to occur where overlapping of the upper and lower circles is shown. Some conditions can develop anywhere in the abdomen. Gynecological conditions causing pain are not indicated. Diseases of the ovaries and tubes usually cause pain in the LLQ or RLQ, and disease of the uterus causes pain in the lower middle area.

Sure, everyone will suffer from abdominal pain at some time, and I've tried to address many common causes in the previous chapters. But what if you're still not confident in your diagnosis? You return to the doctor, and you might be told, "It's just stress." This happens all the time, like with my patient Catherine. Don't settle for a brush-off.

Gallstone Disease

Catherine, who was in her early sixties, enjoyed excellent health except for occasional hunger pains. Then, over the course of several months, she developed stomach pain and bloating, which started in her upper abdomen. Everything she ate made her stomach hurt. She tried not to eat much for breakfast, maybe just a piece of fruit. For lunch, she'd usually eat soup and a roll. For dinner, it was Lean Cuisine or maybe a lamb chop with vegetables. But no matter what she put in her mouth, a short time later she had pain! Remarkably, she hadn't lost weight.

She went for a colonoscopy when the pain first surfaced, which showed diverticulosis (outpouching of the lining of the colon, not a cause for this kind of pain) and two benign precancerous polyps, which were removed. Previously, Catherine had been treated (twice) for a possible *H. pylori* infection and given a trial of Protonix, but it hadn't helped.

She also had undergone an upper endoscopy the previous month, which revealed a small hiatal hernia and some mild redness of the stomach, but no active gastritis (inflammation) and no *H. pylori,* leaving no explanation for the pain. Even so, her doctor started her on over-the-counter Prilosec. While she was on Prilosec, her symptoms somewhat improved. Some days she'd even be completely pain-free. However, the bloating remained and she had persistent nausea.

It was a mystery. Her personal history was good: She didn't drink, and she'd quit smoking ten years before. She wasn't obese and never had been. Her family history was normal. The only slight red flag was a recent trip to Mexico, but her symptoms had started up before she even left for the trip. At this point, after the endoscopy and colonoscopy and a normal blood test, she came to see me.

She did have noticeable bloating throughout her abdomen on examination, as well as tenderness under her breastbone at the top of her abdomen. Because Catherine was better on the Prilosec and her pain seemed to be due to a problem with her stomach or esophagus (dyspepsia, better known as indigestion), I switched her to Nexium to try to eliminate the residual pain. A gastric emptying study showed that the stomach was slow to start emptying but caught up and within two hours would release all its contents. I couldn't really suggest many lifestyle changes—she was already eating a low-fat diet and small meals.

When she returned two months later, she still had symptoms. At that visit she told me that if she ate only soft fruit, like melon or applesauce, she was fine. However, if she ate firm food, she developed pain in her upper abdomen, twenty minutes to one hour after eating. She felt like food "sat" in her stomach, causing loud noises and belly bloat. In the two weeks leading up to our visit, she said, things had only gotten worse and she'd had only one pain-free day. The pain—achy, sharp and burning—now occasionally awakened her from a sound sleep. During her waking hours, she felt like she was constantly expelling gas. Her belly remained bloated, with upper abdominal tenderness across her whole abdomen on exam but no occult (hidden) blood in the stool.

Catherine was clearly worse. Although her blood tests had been normal previously, I repeated them, looking for a clue to her condition. Liver function tests were normal, but the amylase, an enzyme made in the pancreas, was mildly abnormal. Because of her persistent and worsening symptoms and the abnormal amylase, Catherine underwent an abdominal CAT scan to look at the upper abdomen, including the pancreas. The CT scan was normal. An ultrasound of the abdomen showed numerous gallstones and mild thickening of the gallbladder wall, but no fluid around the gallbladder, which suggested that there was no acute inflammation of the gallbladder (cholecystitis). An X-ray study of the small bowel was normal.

Because no other cause could be found for the worsening pain other than the abnormal gallbladder with stones even though her symptoms were atypical, Catherine underwent removal of her gallbladder via laparoscopic surgery. It turned out she had numerous stones, and the gallbladder wall showed chronic as well as active inflammation.

One year after her gallbladder was removed, Catherine remained pain free.

Diagnosing Gallstone Disease

Although gallstone disease is usually easy to diagnose, the symptoms can sometimes be nonspecific, and frequently there are no symptoms at all. Gallstone disease is one of the most common gastrointestinal problems in the United States. In 1994 an estimated 14.2 million women from age twenty to seventy-four had gallbladder disease. The most frequent cause of gallbladder disease is gallstones. They can cause pain and spasm of the duct, inflammation of the gallbladder wall (cholecycstitis) and infection of the bile ducts. In 2004 there were an estimated 1.8 million ambulatory care visits for gallstone disease, with two to three times as many women as men with the diagnosis. There are many clues to making a diagnosis. One is ethnicity. Native Americans have a high incidence of gallstones, with the Pima Indians having an incidence of 70 percent in women over the age of twenty-five. The occurrence of gallstones in women is over one in four for Mexican Americans, one in six for non-Hispanic whites and one in seven for non-Hispanic blacks. Other factors that make gallstones more common are obesity, large and rapid weight loss, pregnancy, age (over forty), a history of diabetes, cirrhosis and use of certain medications, such as oral contraceptives and some antibiotics. Genetics play a role, too. If your mother or sister had gallstones, then you're at increased risk.

The gallbladder sits under the liver in the upper abdomen, toward the right side. Pain from inflammation or irritation due to gallbladder stones in the gallbladder or in the ducts is usually in the upper midline or just to the right of the midline (see Figure 7-1), occurring after meals and usually aggravated by fatty foods. The pain may travel to the back and is often associated with nausea. Most gallstones are discovered incidentally. Over a ten-year period, 22 to 26 percent of gallstones cause symptoms, and only 3 percent cause complications, such as inflammation of the pancreas (pancreatitis), inflammation of the gallbladder wall (acute cholecystitis) or a stone getting stuck in a duct as it escapes the gallbladder.

Any time you have worsening pain or pain that awakens you from sleep, it's time for further evaluation. An abdominal ultrasound is usually the first test ordered. It's better than a CT scan for seeing the gallstones in the gallbladder and usually for evaluating the bile ducts, the tubes through which the bile travels out of the liver and gallbladder, to see if they are dilated. Inflammation of the wall of the gallbladder (cholecystitis) will show up as a thickened or swollen (edematous) area with possibly surrounding fluid. If you have inflammation of the gallbladder, you may have some discomfort when the ultrasound probe is pressed over the gallbladder. An MRCP (an MRI scan with a contrast agent) is an excellent way to visualize the bile ducts as well as the pancreas. A CT scan is a good way to visualize the pancreas, although modification of the CT technique is needed for a careful look at the pancreas. If there is concern that there might be a stone in the bile ducts, an endoscopic study (ERCP) can be done in which a scope is passed through the mouth and then contrast is inserted into the common bile duct to visualize the duct. Stones can be removed this way. A HIDA scan (a nuclear medicine scan in which dye is concentrated in a functioning gallbladder and then excreted in the ducts) is often done to evaluate whether the gallbladder is working or if the duct into it is blocked if there is uncertainty about whether the gallbladder could be causing abdominal pain.

Colorectal Cancer

Ever since Katie Couric underwent a colonoscopy on national TV, we've become much more aware of this disease, which is a good thing. About twenty-four thousand women were expected to die from colon or rectal cancer (CRC) in 2009. Celebrities like Sharon Osbourne and Supreme Court justice Ruth Bader Ginsburg have put a public face on the disease; too often, though, this is thought of as a guy's problem. It isn't. Here's what you, as a woman, need to know:

- Recognize the warning signs of colorectal cancer (blood, abdominal pain, change of stools, weight loss) and seek medical attention if they occur.
- Women (and men) are at increased risk for colorectal cancer if there is a family history of colorectal polyps or colorectal cancer or certain other cancers, like uterine or ovarian; a personal history of uterine or ovarian cancer at a young age; or inflammatory bowel disease.
- Since polyps are the precursor to cancer, colorectal cancer can be prevented in most people if polyps are removed, and it can be cured if caught early. At the first stage, the cure rate is a reassuring 95 percent. Since 1998 the incidence rate of colorectal cancer has been decreasing by 2.2 percent per year for women—perhaps because of better screening and the removal of polyps. The lifetime probability of a woman developing colorectal cancer between 2003 and 2005 was one in twenty (one in eighteen for a man).

Is it necessary for me to get a screening for colon cancer? At what age should I be screened?

Every woman should be screened for colon cancer! An average risk, asymptomatic woman should start screening at age fifty, or at age forty-five if you are African-Americans. However, if your father, mother, child or sibling had colon cancer before the age of sixty, screening should start at age forty or ten years before the age at which the youngest person was diagnosed with the cancer. (So if your mom got colon cancer at age forty-five, you should start screening at thirty-five.) Screening should also start earlier for a person who's had ovarian or uterine cancer before age fifty, and in families with a history of polyps or colon cancer. Screening is done by testing stool for occult (hidden) blood (although many tests are not sensitive) and by visualization of the colon (currently most often by colonoscopy).

Are certain ethnic groups more prone to colon cancer?

Colon cancer tends to strike African-American women more often than it does women from other racial groups; it strikes these women 17 percent more often than white women. On the whole, when discovered, polyps in African-American women tend to be larger than those of white women. In addition, the reported overall five-year survival rate from colorectal cancer is 11 percent lower in African-American women than in white women (55 percent versus 66 percent), likely due to detection of the cancer at a more advanced stage in African-American women. Early detection of colon cancer and removal of precancerous polyps is important for everyone.

What happens during a colonoscopy? It looks painful.

While it's not exactly a walk in the park, you shouldn't be afraid of getting a colonoscopy. A colonoscopy is the endoscopic examination of the large colon and the last part of the small bowel. Here's what happens: We use a fiber-optic camera attached to a flexible tube, which is inserted through the anus. It allows doctors to glimpse any polyps, cancer, inflammation and so forth developing inside the colon. Of course, the colon has to be clear for the procedure to be successful—which means you need to be prepared. Depending on what your doctor tells you, you'll have to drink clear fluids and follow a low-fiber diet for a couple of days. The day before the procedure, you'll be given a laxative to fully clear out the bowel. Most people want anesthesia on the day of the procedure to minimize discomfort and for relaxation. If you get conscious sedation, you'll still be awake for the entire event and able to see your colon on the TV monitor, just like Katie Couric. Deeper anesthesia with propofol is being given by some physicians, which will put you to sleep. You should recover from the sedation in about a half hour to an hour, and you might experience some flatulence after the procedure. (Sorry.) Some people request more anesthesia; some ask to be wide awake and have no anesthesia.

I have heard that a virtual colonoscopy with a CT scan is just as good as a colonoscopy to detect polyps and colon cancer. It seems like a quick and easy test. Should I ask my doctor to do this test instead of the colonoscopy?

Virtual colonoscopy, or CT colography, as it is called, has recently emerged as a way to detect colorectal polyps. It is a CT scan in which specialized techniques are used to visualize polyps. It also requires a trained radiologist for interpreting the X-rays. At this time, if done in an experienced center, studies show that for polyps greater than two-fifths of an inch (ten millimeters) a virtual colonoscopy may be as good as a colonoscopy to detect polyps or colon cancer. For polyps six to nine millimeters in size, virtual colonoscopy may detect only about three in five polyps. How useful it is at detecting even smaller polyps remains to be determined. If a polyp is found, you will still require a colonoscopy to remove it. This happens in about 17 percent of people (about three in twenty). If you thought you could avoid a laxative cleansing before the virtual colonoscopy, sorry, but you're wrong. Your colon has to be just as clean for the virtual colonoscopy as for the colonoscopy. If you require both tests, and a colonoscopy cannot be done on the same day as the virtual colonoscopy, then you will need a second colon cleansing. Because of questions about the accuracy of this test, the unknown requirements for frequency of screening, and other issues that need to be addressed for this to be a widely used procedure, the United States Preventive Services Task Force did not endorse this procedure for routine colorectal cancer screening in its published 2008 report. At this time, insurance does not routinely cover virtual colonoscopy for colorectal cancer screening.

Are there any simple lifestyle modifications that I can make to decrease my risk for colorectal cancer?

Stop smoking, lose weight if you are obese and exercise routinely. In one study women who participated in high-energy exercise several times per week had about half the risk for colon cancer as those women who did not. Be sure to ingest the recommended daily calcium and vitamin D amounts. Long-term use of aspirin has been shown to reduce the risk for colorectal cancer (CRC), but because of potential bleeding as a side effect, it is not recommended for prevention of CRC.

I'm twenty-four years old and have had bloody diarrhea for the last three weeks. I'm so scared that I could have cancer and haven't told anyone. What should I do? I'm almost afraid to find out what's wrong.

First of all, it is very unlikely that you have colon cancer. Colon cancer is more likely to occur in women fifty and older. However, just because you're young, you shouldn't ignore the bleeding. Colon cancer can occur in young women. Do you have a strong family history of polyps or cancer? Have you had inflammatory bowel disease for more than ten years?

As for possibilities, polyps can sometimes bleed, and some types of polyps can be present even in a young person. You also might have an infection. Amoeba infections, *Clostridia difficile* infections that occur after antibiotic use, and occasionally some bacterial infections can last as long as three weeks. However, the most likely cause of your bleeding is colitis, inflammation of the colon or large bowel. This condition first occurs most commonly in twenty- and thirtysomethings.

You need to overcome your fear and see your doctor. In the long run, it's much better to know what's wrong and to be treated than to ignore the problem and risk the probability that it will get worse. The first thing that your doctor is likely to do is to examine you, get routine blood tests to make sure that you aren't anemic and to see if you have indications of inflammation, and obtain stool cultures. Then she is likely to recommend a colonoscopy, which will visualize your colon. Areas that look inflamed or abnormal can be sampled, and if you have polyps, they can be removed.

Other Issues

My sister had a really sharp pain low down in her right side and a fever. She went to the hospital and they told her that they thought she had appendicitis and would need surgery. But when they did a CT scan, they told her that she had diverticulitis and should be treated with antibiotics. Why the confusion?

Diverticula are outpouchings of the bowel. They probably form when the bowel wall doesn't contract equally well throughout, and so a small area ends up being pocketed out, in the same way balloon figures are made. These diverticula are more common on the left side of the colon, particularly in the sigmoid colon, but can occur anywhere in the colon and even in the small intestine. The existence of diverticula is called *diverticulosis.* Inflammation of a diverticulum is called *diverticulitis.* Diverticulitis on the right side of the abdomen could be due to a twisty sigmoid colon that resides partly over in the right side of the belly or to a diverticulum in the ascending colon or cecum (the right side of the colon) that is inflamed.

The appendix, meanwhile, is like a hollow tail off the cecum on the right side of the colon. Inflammation of the appendix usually presents as pain in the mid-abdomen that migrates down on the right side. This type of pain and pain from inflammation of a diverticulum often can't be distinguished without further imaging. Disease of the right ovary or tubes or inflammation of the last part of the small intestine (terminal ileum) due to infection or inflammatory bowel disease (Crohn's disease) can also cause identical symptoms. (Crohn's disease can cause inflammation in the small intestine, colon or both.) Whereas acute appendicitis is usually treated with surgery, the first episode of acute diverticulitis is often treated with antibiotics. For diverticulitis, surgery is usually done if there is a complication—blockage, an abscess, a hole in the bowel or a connection from the diverticulum to another organ— or repeated episodes of diverticulitis in the same location. The CT scan is an excellent test for diagnosing the cause of the pain. Occasionally, colorectal cancer can masquerade as diverticulitis. Therefore, I usually recommend a colonoscopy after the diverticulitis resolves.

What is colitis?

Colitis is inflammation of the colon. It can be caused by an infection or by other conditions that cause inflammation. The best known and most common type of noninfectious colitis is inflammatory bowel disease (IBD). The two primary types of IBD are ulcerative colitis (slightly more common in men) and Crohn's disease (slightly more common in women). The symptoms of ulcerative colitis or Crohn's colitis may be somewhat different. We don't know the cause of these conditions, although we are getting closer in our understanding. They appear to be caused by an interaction of a genetic predisposition, normal and abnormal bacteria in the gut, inflammatory substances released by cells in our body that cause both bowel and more generalized inflammation, and environmental factors. (One example of an environmental factor is smoking, which decreases the risk of ulcerative colitis, while it increases the risk for Crohn's disease.)

Ulcerative colitis consists of inflammation of the lining of the large bowel starting with the rectum and spreading upward. It may involve only the rectum (a condition called proctitis), or it may involve all or a part of the colon. The inflammation is continuous and does not skip around. It usually shows up as bleeding, frequently accompanied by mucus and often with an urge to have a bowel movement, even if you pass only mucus or blood. Often there is abdominal cramping, diarrhea and sometimes even a fever. If you have only proctitis (inflammation of the rectum), it often will remain in the rectum and not involve any more of the colon. However, 30–50 percent of the time, it becomes more extensive.

Crohn's disease-caused colitis may also result in bleeding. However, Crohn's disease usually doesn't have the same continuous involvement of the colon. It can involve only patches in the colon, leaving normal areas in between the affected areas. If the colitis is due to Crohn's disease, it is very important to make sure that the inflammation does not involve the small intestine. In Crohn's disease if you have involvement of the colon, there is about a seven out of ten chance that there is also inflammation of the small intestine or elsewhere. In one out of three women with Crohn's disease, inflammation around the anus can occur. This can cause pain and bleeding similar to hemorrhoids and to anal fissures from constipation. At the time of a colonoscopy, when your doctor is examining the large bowel, she will try to enter into the last part of the small intestine (terminal ileum) to take a peek to see if there is inflammation. She will also take samples of tissue in the colon. This will help determine the type of colitis that you have. If it looks like you may have Crohn's disease, then she may want to do further testing to evaluate the small intestine.

These tests include a small bowel follow-through barium test, an abdominal CT or MRI scan with a special technique to look at the small intestine (enterography) and an enteric capsule study. With a small bowel follow-through, you swallow thick, white, chalky-tasting barium, which is followed through the small intestine with fluoroscopy while numerous X-rays are taken. This test is now being replaced in many places with either an abdominal CT scan or MRI scan, in which a patient swallows an amount of diluted barium that allows these scans to specifically see the outline of the small intestine. With this technique, inflammation outside the bowel wall, as well as inside the bowel, can be assessed. This helps determine if there is active disease, as well as inactive disease. It can also reveal abscesses (collections of pus) outside of the bowel. The enteric capsule study involves swallowing a capsule that takes pictures inside your bowel and transmits them to a recorder that you wear for a twenty-four-hour period. The capsule actually sees the ulcerations and inflammation, rather than just suggesting a problem based on the appearance of the contrast outlining the wall. Note: The capsule can't be used if there is even a chance of narrowing in the bowel, as the capsule might get stuck. Your doctor will determine which test is right for you.

If you do have colitis, your doctor will discuss its possible treatments. For inflammation limited to the left side of the colon, the treatment is often local, with a suppository or an enema consisting of a salicylate (a drug related to aspirin) called mesalamine. A steroid enema might also be used. The enema is a small amount of liquid that is squeezed into the rectum (like a douche into the vagina), usually at night. It likely will remain there overnight. Unlike the enemas that we use for constipation, the mesalamine and steroid enemas don't usually stimulate you to have a bowel movement. Mild or moderate disease extending higher than the reach of enemas is often treated with tablets of mesalamine with or without local treatment. Crohn's disease may be treated with antibiotics. Recent studies have examined the use of probiotics in IBD, and there are some studies suggesting their utility. Traditionally, the order of therapy for mild to moderate IBD has started with mesalamine, progressing to corticosteroids, such as prednisone, which is a strong anti-inflammatory medication, and/or to immunosuppressant therapy, like azathioprine (Imuran), 6-mercaptopurine (Purinethol) or methotrexate (Crohn's disease), then to the biologic therapies. Currently the anti-tumor necrosis factor (anti-TNF) medications infliximab (Remicade), adalimumab (Humira) and certolizumab pegol (Cimzia) are the most commonly used biologics (substances made from living organisms or their products, used for treatment of disease). Steroids act rapidly but are not used long term to keep the symptoms at bay. The immunosuppressant therapy takes six weeks to six months to work but has a good long-term effect on the disease. The biologics can both rapidly treat the inflammatory bowel disease and can keep the active disease from coming back. Severe ulcerative colitis or Crohn's disease is treated with steroids, biologics or immunosuppressive drugs, such as cyclosporine (for ulcerative colitis), that act relatively quickly. Recently, there has been a debate among specialists about whether to use immunosuppressant therapy early in the treatment of mild to moderate Crohn's disease rather than *after* other therapies have failed. The rationale for this approach is that the anti-TNF medications may work better in people who have never had steroids; there is an increased risk of infection if a person starts the anti-TNF therapy while on steroids, and steroids have a large number of unpleasant, visible side effects, such as a balloonface (like a chipmunk with nuts in his cheeks), acne, weight gain and irritability.

There are major possible side effects with the anti-TNF therapies also, although these are different from those of the steroids. These include an increase in infections, especially tuberculosis and fungus infections, or activation of hepatitis; an increase in tumors, including lymphoma (although small to date); and, more commonly in women, the onset of SLE (systemic lupus erythematosis) and an effect on the nerves, similar to multiple sclerosis.

My approach to a patient's IBD is individualized and changes as new data become available. At the time of this writing, I am in favor of the bottom-up approach, utilizing the anti-TNF therapies after the other treatments. However, I try to stop steroids, if used, as soon as possible in order to minimize side effects. Because azathioprine or 6-mercaptopurine take a long time to start having an effect on IBD, a patient's acute symptoms must already be controlled with another medication before using either of these two medications. It's not a quick fix. The anti-TNF and steroid therapies work much faster. Monitoring of the liver function and the white and red cells and platelets should continue when on azathioprine or 6-mercaptopurine.

I don't have bleeding, only abdominal pain. After testing, my doctor said that I have Crohn's disease. I thought that I would have to be bleeding to have Crohn's disease. Can she be right?

In about one out of three people with Crohn's disease, the colon is not involved, and then visible bleeding is uncommon. People with Crohn's disease of the small intestine often just have belly pain. Other symptoms can be diarrhea or constipation, weight loss, fever, fatigue or loss of appetite. In some people, the small intestine or colon can become narrowed, and the stool may have a hard time getting through. This can cause constipation or simply pain. A fullness or mass in the belly can occur due to the inflammation, and some women discover this problem even before they go to a physician.

Ever since I was diagnosed with IBD, my sex life has gone downhill. I feel so isolated—and celibate!

Don't feel alone. Women with inflammatory bowel disease have several problems that can affect their sex lives. More than two out of five female sufferers report that IBD has an adverse effect on their life enjoyment, about one in five women report being dissatisfied by their body image and one in four women report having infrequent or no sexual intercourse, either due to pain or poor self-image. If you have belly pain or dread suddenly going to the bathroom in the throes of passion, your desire for sexual intercourse is likely the last thing on your mind. Not to mention that putting an enema into your rectum at night isn't exactly a mood setter.

There are ways to ease your troubles. Make sure that you take all your medications. If a food bothers you and causes pain, try to avoid it, especially at dinner, when the effect can persist—killing your sexual appetite at bedtime. Try to pass gas and empty your bowel before going to bed or before sex. Delay inserting your enema or suppository until after sex in the evening or delay the sex until the morning.

Interestingly, in a study of fifty women with Crohn's disease, 24 percent had infrequent or absent sexual intercourse, compared with 4 percent in a control group. Pain occurring during intercourse was the reason cited in 60 percent of these women with Crohn's disease and 38 percent of women with ulcerative colitis complaining of infrequent intercourse. Often this is due to inflammation around the rectum, anus and vagina. The contribution of vaginal dryness was not examined. Try using a vaginal lubricant and see if that helps. After women underwent surgery for ulcerative colitis and had their colons removed and a rectum-like reservoir for stool made from the ileum, the rate of preoperative sexual dysfunction of 73 percent had markedly decreased to 25 percent twelve months post-surgery.

Of course, comfort with your own body image contributes to your desire to have sex. If you have had previous surgery, this might be particularly distressing. How do you explain that your waste comes out through an opening (called an ostomy or stoma) and collects in a bag on your belly? How can you forget about it and make it attractive? If you have an ileostomy or colostomy, empty your bag if you think that sexual intercourse may be in the offing. You can even purchase attractive covers for the stoma bag online, or if you are handy with a needle and thread or a sewing machine, make your own. Be sure to discuss this with a potential partner ahead of time. A stoma does not have to be a deterrent to a happy sex life.

Side effects from the corticosteroids, like weight gain, acne or disfiguring lines on the belly (called striae), might make you feel less attractive and not like your "real" self. If this is a problem, discuss it with your physician and see if another medication is possible for treating your disease. Because scars can also negatively impact one's body image, discuss with your surgeon ahead of time whether you might be a candidate for a laparoscopically assisted surgery that leaves smaller scars.

I'm thinking of getting pregnant, but I was diagnosed with IBD. Is this a bad idea?

Not if you're in remission. In remission, women with IBD are likely to have uncomplicated pregnancies without any risk of miscarriage. (There is an increased risk that you may give birth before thirty-seven weeks, so just make sure you don't wait until the last minute to prepare the baby's room!) Be sure to discuss your medications with your doctor to understand any potential risk to your baby. Most medications are safe. (See Chapter 8.) While having IBD is definitely no reason not to become pregnant, you need to work with your doctor to achieve and maintain remission before getting busy. Active Crohn's disease is associated with complications and will likely remain active during your pregnancy.

REMEMBER TO SEE YOUR DOCTOR IF:
- You have bleeding from your rectum.
- You have abdominal pain that is severe and occurs suddenly.
- You have chronic abdominal pain that is awakening you from sleep or pain for which you have no satisfactory explanation.
- You have unexplained weight loss.
- You have a sudden inability to move your bowels.
- You have a sudden change in your stools that persists more than a couple of weeks.

Chapter 8
Nine Months of This?
Minimizing Stomach Problems
During Pregnancy

*Because they've either conveniently forgotten with time
or they're trying to be supportive, most mothers won't tell
you how hard pregnancy (and then childbirth) can be.
Let me tell you, it is. It's brutal sometimes.*
—Jenny McCarthy

A h, pregnancy—most of us have either been there, done that or would like to at some point. Pregnancy itself is certainly an adjustment under the best of circumstances. We have to cope with major physical changes. So long, skinny jeans! Nice knowing you, crash diet! It's the one time in a woman's life when she absolutely *must* embrace weight gain and healthy eating.

To complicate things, pregnancy's a time when constipation, heartburn, nausea and bloating become exacerbated. In this chapter I'll try to help you make the longest nine months of your life memorable—not excruciating—from a GI standpoint.

During my pregnancy I was very sick and had to have an emergency CT scan of my pelvis when I was twenty-four weeks along. I'm very worried that my baby will be affected. Will my baby be at risk for birth defects or cancer?

Your baby is at most risk from X-rays when the fertilized egg is implanting (taking up residence) in the uterus and then between two and twenty weeks. Between two and twenty weeks, the fetus is most at risk for malformations from radiation. However, anything below a threshold dose should be okay. It's tough to determine exactly what the threshold dose is, though. It is thought to be between five and fifteen rads. (A rad is a measure used for a dose of ionizing radiation.) The dose of radiation from a pelvis CT scan is about one to five rads. In practice, no increase in congenital birth defects in the fetus from one pelvis CT scan has been seen. The subsequent cancer risk to your child is very low and only slightly increased after radiation. The risk of a child dying from childhood cancer will increase a tiny bit after radiation, from one in two thousand (without radiation) to two in two thousand (after exposure to four rads of radiation). This risk is greater when the radiation exposure is in the first trimester and barely increases when the exposure is in the second or third trimester.

What is the preferred type of imaging in pregnancy?

That depends on what the doctor suspects is wrong. An ultrasound is safe at any time. If you have pain in the upper right side, the doctor may be worried about gallstones or stones in your bile ducts, the tubes leading out of the liver and gallbladder. An ultrasound is the first course of action when examining the gallbladder and the bile ducts. Inflammation of the appendix can be very tricky to diagnose in pregnancy because the appendix's usual location, low in the right side of the abdomen, attached to the last part of the colon (cecum), often changes. As the baby grows, the uterus displaces the cecum and appendix upward. Therefore, appendicitis can be incorrectly diagnosed as another problem. If it is suspected, an ultrasound may be very helpful in the diagnosis of appendicitis but may not be sensitive enough.

An MRI scan is the more sensitive choice for diagnosing appendicitis or for evaluating other problems in the abdomen. But it's usually not the first choice for evaluation, because it is expensive. MRIs are felt to be safe in pregnancy, but the data are limited—there's no clear difference in the risk of an MRI to the fetus between the three trimesters. But you should note that, if "very" pregnant, you might be too large to fit under the scanner. The intravenous contrast gadolinium (used to examine the blood vessels during the MRI scan) should *not* be used in pregnancy as it may create a possible kidney problem in the mother or fetus. On the other hand, after you deliver, it would appear that it is safe to continue breast-feeding after gadolinium exposure, although in the past it was recommended that a woman stop breast-feeding for twenty-four hours after receiving contrast. A single CT scan during pregnancy is probably safe and could be done if an MRI is not available or cannot be done. Shielding of the fetus in the pelvis, and doing a limited study to examine the area of interest, is the best approach.

I began bleeding from my rectum in the sixth month of my pregnancy. My doctor wants me to have a colonoscopy to look in my colon to see what's going on. Is it safe?

Yes. Colonoscopies and upper endoscopies are safe in pregnancy. However, we try to delay them until after delivery if at all possible, since it's thought they *could* induce premature labor. Bear in mind that during the endoscopic examination most obstetricians and gastroenterologists will want to monitor your baby's heartbeat. As for medications given for sedation, they're usually modified to become pregnancy proof. Remember, though, that colonoscopies are technically more difficult to do as the growing uterus displaces the colon upward.

I'd do an endoscopic exam if I felt it was crucial to evaluate a symptom that could be treated during pregnancy. Rectal bleeding is one such symptom. There are several potential causes for bleeding: Something as simple as hemorrhoids may be the cause and can be treated locally and by keeping your stools soft; if there's a lot of bleeding, the hemorrhoids can be banded. You could have inflammation—colitis or proctitis. This can be treated safely with medication during pregnancy and treatment for these shouldn't be delayed until you deliver. Polyps are another possibility, and they can be removed during pregnancy. Don't worry. It's rare for a woman to develop cancer during her pregnancy, but it can happen, and this needs to be ruled out, which a colonoscopy can do.

I'm contemplating getting pregnant, but I take lots of GI medications. How do I know if the medications are safe for my unborn baby?

In the United States the FDA assigns a risk category to the drugs that it approves based on the level of risk a drug poses to a fetus. The ratings are based on clinical studies done in pregnant women, clinical studies done in pregnant animals and/or post-marketing surveillance of the outcome of pregnancies in women taking the drug. Very few controlled studies are done in pregnant women.

Therefore, there are limited data on the safety of many medications used to treat gastrointestinal diseases and other diseases in pregnancy. Evaluating drug safety by doing studies in animals does not necessarily mean that those results would correspond to results in pregnant women, but often animal studies are the best information available. The FDA is currently re-evaluating the reporting of drug safety in pregnancy. Below are the current FDA categories for drug safety in pregnancy.

Category A and B drugs are considered safe drugs for the pregnant woman. Category C drugs are those for which there are no studies in humans available or which may pose a risk to an animal's fetus. These drugs are used if the benefit is thought to be greater than the risk.

Category D drugs are those for which there is definite evidence of risk to the unborn child. However, the benefits to the mother and fetus may out-weigh the risk to the fetus. In other words, if you have an illness that is kept in check by a drug in this category, and if you stopped the drug, you may get sick. If you got sick, a healthy pregnancy outcome may be less likely than if you had continued taking the drug.

Category X drugs should *not* be used in pregnancy. The risk of the drugs clearly outweighs any benefit from taking them. One notorious drug is thalidomide, which caused severe malformations in human fetuses, such as missing arms or legs. Many women took this drug for anxiety in the 1950s.

The most important thing that you can do to assure your baby's safety is to discuss your medications with your obstetrician, gastroenterologist and primary care physician *before* you get pregnant. If there are medications that could potentially have an ill effect on your baby, then it might be possible to switch to a medication with similar benefits that has a better safety profile in pregnancy. See Table 8-1 for the pregnancy categories of common gastrointestinal drugs.

I developed heartburn after only two months of pregnancy. I expected to have heartburn in the third trimester. Why did it start so early? When will it stop? And what can I do about it?

Heartburn is a common complaint in pregnancy. It affects 40 to 80 percent of pregnant women. And it can occur anytime in each of the three trimesters. You might think that heartburn would be worse in the third trimester due to weight gain and an enlarging uterus, but not necessarily. Pressure in the lower esophageal sphincter (at the bottom of the esophagus, just above the stomach) decreases as a result of the effects of the hormones progesterone and estrogen, although this usually happens in the second and third trimesters. Delayed emptying in the stomach and abnormal motility in the esophagus may contribute to the reflux.

Not surprisingly, if you had episodes of reflux before you were pregnant, you are more likely to have frequent heartburn in the first trimester. But, take heart: the heartburn usually resolves within a few days of delivery if it is newly associated with the pregnancy. However, after pregnancy, some women might find that they continue to suffer from heartburn.

The initial treatment for heartburn and reflux in pregnancy is the lifestyle changes described in Chapter 6. Avoid the trigger foods that cause heartburn. Don't eat within three to four hours of bedtime and elevate the head of your bed when sleeping. If lifestyle changes don't cure the problem, then I recommend trying antacids. These are safe in pregnancy. If antacids are not working well enough, I then recommend an H2RA (H2 receptor antagonist) for treatment. Many of these are available over the counter and include ranitidine (Zantac), famotidine (Pepcid), cimetidine (Tagamet) and nizatidine (Axid). They are all category B. If maximum doses of these drugs don't work, I would recommend a proton pump inhibitor (PPI), all of which are category B, except for omeprazole.

I don't have morning sickness—I have all-day sickness! My nausea can take over at any random time. It's ruining my social life and making work almost impossible. When will this end?

Nausea and vomiting in pregnancy afflict 70 to 80 percent of pregnant women—some more frequently than others. And morning sickness is a misnomer—queasiness can strike at any time. In most women, nausea and vomiting are better by the end of the first trimester. By week twenty-two (a little over halfway through pregnancy), over 90 percent of women no longer have these problems. For an unfortunate few, the nausea and vomiting can be persistent. This constant vomiting is called hyperemesis gravidarum.

There are treatments. For simple nausea and vomiting, try to modify your eating habits. Make sure you keep hydrated with fluids. Try to drink sport drinks, soups and broths. If you can handle these, try adding starches and protein to your diet. Avoid fatty foods. Eating small amounts of food throughout the day may also help.

If this doesn't help, vitamin B$_6$ or ginger might do the trick. I recommend 10 to 25 mg of vitamin B$_6$ three times per day. Ginger can be taken as a capsule or a syrup at 1000–1500 mg in three to four divided doses.

For the unrelenting nausea and vomiting of hyperemesis gravidarum (and sometimes for the nausea and vomiting in the first trimester of pregnancy), other medications may be needed. One of these medications is metoclopramide (Reglan), a category B drug that is frequently used during pregnancy in the United States. (The FDA recommends that its use be limited to three months unless needed for severe symptoms.) Some women have found relief through acupuncture or acupressure, including wearing wrist bands that are used for motion sickness. Studies have been inconclusive as to the effectiveness of these treatments, but it's certainly worth a try.

Will my celiac disease interfere with my pregnancy?

If you have celiac disease, it is important to rein in your symptoms before you try to get pregnant. Your fertility and the outcome of your pregnancy should be normal if your disease is under control. However, if your disease is *not* under control, the baby may be underweight; it may be delivered early, before thirty-seven weeks; and/or a Cesarean delivery might be necessary. Many of these adverse outcomes of the pregnancy may be due to malnutrition or a low level of nutrients in the body.

I have irritable bowel syndrome and have done well on my medications and am essentially pain free. Can I safely remain on my medications during pregnancy? Will my IBS symptoms get worse in pregnancy?

There are no data to show that IBS either improves or worsens during pregnancy. However, treatment with some of the medications used for IBS may pose a risk to the fetus. You should discuss with your gastroenterologist and obstetrician whether you should remain on your current medications or whether your medications should be changed prior to conception.

Lindsay's history of IBS and her special concerns before and during her two pregnancies are typical of many pregnant women with this condition. I met Lindsay, then thirty-five, when she needed a second opinion regarding her irritable bowel syndrome. A few years prior to being seen, Lindsay started gaining weight, so she began swimming and working out. During and after exercise, she developed a jabbing pain on the right side of her abdomen that initially seemed to respond to nonsteroidal anti-inflammatory medications. However, a short time later the pain seemed to take on a life of its own—she experienced a stabbing pain and bloating when she ate, sometimes severe enough to wake her up. She required laxatives (Correctol and magnesium citrate) to have a bowel movement every other day. Because of the pain, she underwent an abdominal ultrasound, a colonoscopy, a small bowel X-ray and even a laparoscopy, all of which were normal. Then she was diagnosed with irritable bowel syndrome. She was treated with Effexor (venlafaxine, a serotonin/norepinephrine receptor inhibitor) and did well for quite some time.

Lindsay wanted to get pregnant, and she didn't want to take venlafaxine during pregnancy. Before she came to see me, she'd begun to taper off her medication. Gradually, a twinge of pain re-appeared in her lower right side. At dinnertime she became distended. For the IBS with constipation she was started on tegaserod (Zelnorm, a stimulator of a serotonin receptor in the gut), which relieved her bloating and pain. (This medication is no longer available on the market.) The constipation was relieved by MiraLAX.

After seeing me, she successfully completed a slow taper of her venlafaxine, and when seen two months later, she remained pain free. When Lindsay was seen ten months after her initial appointment, she was fifteen weeks pregnant. After becoming pregnant, Lindsay went off tegaserod and MiraLAX and took only a stool softener. But by her twelfth week of pregnancy, her stools had become thinner and had decreased to every other day. Therefore, Lindsay restarted the MiraLAX. Her right-sided pain, which felt like a shock, re-emerged, occurring off and on, and even precipitated a panic attack. The pain occurred after eating or drinking, even when she drank water, and was accompanied by gas rumbling in her abdomen and marked bloating. The MiraLAX helped her pass the gas and stool. She also started enteric-coated peppermint oil capsules before eating, which helped her GI symptoms.

The pregnancy continued to progress well. By twenty weeks Lindsay had gained fourteen pounds and the nausea and every-other-day vomiting that plagued her first trimester had stopped. Her appetite increased; in fact, she was hungry all the time. The enteric-coated peppermint oil capsules and MiraLAX, taken periodically, continued to help her gas and daily bowel movements. She still felt intermittent abdominal pain, though, which caused her severe anxiety and panic attacks. For this reason, I started Lindsay on dicyclomine (Bentyl, an antispasmodic), 10 mg three times per day. Furthermore, her primary care doctor started her on a low dose of citalopram (Celexa), a category C drug similar to the venlafaxine. A new upper right-sided pain, which she described as "like contractions," developed under Lindsay's ribs at thirty-four weeks. Because gallstone symptoms are commonly in this location, an abdominal ultrasound was done; thankfully, it was normal. However, it did determine that Lindsay was very constipated; production of stool helped relieve the pain. By thirty-four weeks, Lindsay had gained a total of twenty pounds. Subsequently, at full term, Lindsay delivered a healthy baby by C-section, done because her labor was very slow to progress.

One month after giving birth, Lindsay continued to have a lower right stinging or pulling abdominal pain. Her bowel movements now occurred once or twice per day with two capfuls of polyethylene glycol 3350 (MiraLAX). She restarted tegaserod twice a day because it had helped her right lower quadrant pain and constipation in the past, and she continued on the dicyclomine for the abdominal spasm and the citalopram (Celexa) for continued anxiety. Because she was sure that there was something serious going on in her right side and had persistent tenderness, I performed a colonoscopy. This showed no inflammation.

By four months postpartum, all of Lindsay's abdominal pain had disappeared, except for one episode of pain and nausea after eating out at a restaurant. Her bowel movements occurred daily on two scoops of MiraLAX and the tegaserod. She continued on the citalopram for the anxiety but rarely needed the dicyclomine, which she took for abdominal pain.

I saw Lindsay again eight months after she gave birth. She told me she was pregnant again—seventeen weeks along. She'd stopped the tegaserod as soon as she knew she was pregnant. She decreased her MiraLAX to one scoop per day due to vomiting, and her stools remained firm, eliminated once a day. This time she continued on the citalopram due to her anxiety. Overall, her symptoms during this pregnancy were much better than the previous one. Her IBS did not act up. Her abdominal pain was mild, occurring on the right side of her abdomen with her bowel movements. She continued to work full-time throughout her pregnancy.

The pain re-emerged in the right lower part of her belly at twenty-two weeks of pregnancy, one week before it had occurred in the previous pregnancy. The pain would shoot up her right side to just under the ribs. It zapped her and then stopped, lasting for just a moment. She thought that it felt like it was a "nerve pain." The pain lasted for a few weeks, then disappeared. She used the dicyclomine when needed for the pain. When the pain disappeared, she stopped the dicyclomine and she continued to be pain free with a daily bowel movement, now with two capfuls of polyethylene glycol. She still had gas and bloating after eating, however, and her anxiety continued even on the citalopram.

Her second pregnancy progressed more or less uneventfully, despite the sporadic episodes of pain. None of the severe pain that she had experienced in her first pregnancy materialized in her second. She felt that a big part of it was the fact that she continued on the citalopram. Only rarely did she need or take a dicyclomine. The MiraLAX kept her stools regular.

Ultimately, Lindsay delivered a healthy, full-term baby by C-section. Postpartum she continued to do well. One year after delivery she remained pain free, with daily bowel movements, on only the citalopram and MiraLAX powder.

Lindsay's two pregnancies were different in regard to the severity of her IBS symptoms. In the first pregnancy her IBS symptoms were severe. In the second they were mild and intermittent. The only difference between the pregnancies in terms of medication was a slightly higher dose of the citalopram in the second, which was unlikely to have made the difference. You can't predict if IBS will be a nonissue or a problem in pregnancy. Of course, when pain does occur in pregnancy, you can't just assume that it's IBS. Other potential problems have to be eliminated as a cause.

Are venlafaxine (Effexor), citalapram (celexa) and polyethylene glycol safe to use in pregnancy?

Venlafaxine and citalapram are category C medications. First trimester use has not been shown to result in increased malformations in newborns. However, third trimester use of SSRI medications (including venlafaxine and citalapran) has been shown to cause behavioral changes in some newborns and possibly even depression of breathing and seizures. These changes are usually present within two days after birth and resolve after three days. Therefore, it is a good idea to try to stop this medication during pregnancy. In cases when the medication is needed to prevent severe depression or anxiety, the benefit of taking it could outweigh the risk of stopping it. Severe depression in pregnancy could potentially adversely affect the pregnancy. Polyethylene glycol (GlycoLax, MiraLAX, GoLYTELY, etc.) is a category C drug and is safe to use in pregnancy. See Table 8-1 for more detail on which drugs are safe in pregnancy, and consult with your doctor for the best treatment for your symptoms.

TABLE 8-1: DRUGS FOR GASTROINTESTINAL PROBLEMS IN PREGNANCY

Nausea and Vomiting in Pregnancy

Drug	Pregnancy Category	Usual Dosage	Additional Comments
Vitamin B$_6$	A	10–25 mg 3 times/day	
Prochlorperazine (Compazine)	C	5–10 mg 3 times/day	Unproven safety but widely used.
Promethazine (Phenergan)	C	12.5–25 mg 4 times/day	Unproven safety but widely used.
Metoclopramide (Reglan)	B	10–20 mg 4 times/day	Appears safe for use short term. Try to limit daily use to less than 3 months. Can cause problems with nerves.
Ondansetron (Zofran)	B	8 mg 3 times/day	Appears safe based on limited data.

Gastroesophageal Reflux Disease (GERD) and Peptic Ulcer Disease (PUD)

Drug	Pregnancy Category	Usual Dosage	Additional Comments
Magnesium & aluminum hydroxide	B	2–4 tsp as needed	Magnesium-containing antacids should be avoided near term.
Sucralfate	B	1 g one hour before meals	
H$_2$ blockers	B	Follow directions based on retail brand	
Proton pump inhibitors (PPI)	B/C	Follow directions based on retail brand	All PPIs, except omeprazole (Prilosec, a category C), are classified as category B.
Calcium carbonate (Tums)	Not classified		Generally considered safe.

Irritable Bowel Syndrome (IBS)

Drug	Pregnancy Category	Usual Dosage	Additional Comments
Loperamide (Imodium)	B	2–4 mg after each unformed stool or 1–4 mg as needed up to 4 times/day	Antidiarrheal agent of choice during pregnancy.
Diphenoxylate/ Atropine sulfate (Lomotil)	C	1–2 tablets 4 times/day	Should be avoided during pregnancy.
Dicyclomine (Bentyl)	B	10–20 mg 4 times/day	Reserved for women with bad symptoms.
Hyoscyamine (Levsin)	C	0.125–0.25 mg 4 times/day	Reserved for women with bad symptoms.
SSRIs (Citalopram, fluoxetine, sertraline, venlafaxine, duloxetine, paroxetine)	C/D	Follow directions based on retail brand	No proven increased risk of congenital malformations except paroxetine, where risk increased in many studies. Use in 3rd trimester associated with withdrawal symptoms and behavioral and neurological symptoms shortly after birth and lasting a few days.
Tricyclic antidepressants amitriptyline (Elavil) nortriptyline, desipramine	C/D	Follow directions based on retail brand	Questionable safety in pregnancy; use should be limited to the severely symptomatic.

Constipation

Drug	Pregnancy Category	Usual Dosage	Additional Comments
Magnesium citrate	A		May need to limit use near term.
Polyethylene glycol (MiraLAX, GlycoLax)	C	17 g with 8 oz water daily	Limited data in women. Considered safe.
Sorbitol	B	2–10 tbsp single dose	Use often limited by cramping and bloating.

Drug	Pregnancy Category	Usual Dosage	Additional Comments
Lactulose	B	1–2 tbsp 1–3 times/day	Use often limited by cramping and bloating.
Senna (Senokot)	C	2 tablets (17 mg) daily	Safe and effective in pregnancy.
Bisacodyl (Dulcolax)	B	10–15 mg	Use limited due to cramping.

Inflammatory Bowel Disease			
Drug	Pregnancy Category	Usual Dosage	Additional Comments
5-ASA (Azulfidine, Asacol, Pentasa, Colazal, Rowasa, Canasa)	B	Follow directions based on retail brand	Can be used safely in oral and topical forms.
Corticosteroids Prednisone	B	variable	
Budesonide	C	9 mg daily	Likely safe but no controlled studies available in pregnant women.
Infliximab	B	5–10 mg/kg IV	Appears safe based on limited data.
6-Mercaptopurine/ Azathioprine	D	Starting at 50 mg	Use is justified if patient has active disease not responding to other medications. Possibly, minimal cause of birth defects in human fetuses.
Drug	Pregnancy Category	Usual Dosage	Additional Comments
Metronidazole (Flagyl)	B	750–1500 mg in 3–4 divided doses per day	Appears safe during pregnancy but use is limited to second and third trimesters.
Fluoroquinolones (Ciprofloxacin), Levofloxacin	C	Follow directions based on retail brand	Limited safety data; usually avoided during pregnancy due to possible effects on collagen development in the joints.

Summary of commonly used drugs for various gastrointestinal ailments during pregnancy. Modified from C. Thukral and J. L. Wolf "Therapy Insight: Drugs for Gastrointestinal Disorders in Pregnant Women." *Nature Clinical Practice Gastroenterology & Hepatology* 3 (2006): 256-266.

How can I tolerate my vitamins with iron in them during pregnancy? Won't they cause constipation?

Constipation often occurs in pregnancy, with the exact incidence unknown. The literature reports the incidence varying from as low as 5 percent to as high as 51 percent, depending on how constipation is defined. What role iron or prenatal vitamins alone play in causing constipation is unknown. Iron and other minerals can cause the stool to be hard or less frequent. If you need the iron to maintain your red blood count, then it is important to try to take it. To help tolerate iron, increase the fiber in your diet, take stool softeners with the iron, or take polyethylene glycol powder—MiraLAX or GlycoLax—as you would to treat constipation. The same fiber supplements discussed in Chapter 5 could be used safely in pregnancy. Indigestible sugars, lactulose or sorbitol, could be used if necessary, but they often cause bloating and gas. Senna should be safe. However, most women would only need to take this compound sporadically. Bisacodyl and magnesium-containing laxatives are also likely safe.

I tend to have loose bowels. What can I take during pregnancy to keep my bowel movements in check?

Diet is the first course of action. Ask yourself: Am I lactose intolerant? Am I drinking more milk or eating more dairy products during my pregnancy? If yes, try cutting back on or cutting out your dairy products for a week and see what happens. If you do find that the lactose is bothering you, you could try Lactaid milk, soy milk, rice milk or almond milk. Make sure to take enough calcium by adding a supplement if necessary. Try to determine if other things that you might be eating could be causing loose or frequent stools, such as fructose, gluten or a specific food. If diet isn't the answer, it's okay to take loperamide (Imodium), a category B drug. Try taking just a small amount of it. For IBS, sometimes just a half to one 2 mg tablet may be enough. Stay away from Pepto-Bismol, as there are some reports of harm to the mother and baby.

REMEMBER:

- It's likely that you'll deal with gastrointestinal symptoms during pregnancy.

- Almost every woman has nausea and most have vomiting in pregnancy.
- Heartburn and reflux commonly occur and can be a nuisance, but they're usually easily treated.
- Constipation is also common but can often be treated easily.

Chapter 9

Eating Your Way to Health

"The second day of a diet is always easier than the first.
By the second day you're off it."
—Jackie Gleason

Most of us are lucky enough to choose what we eat. We can make healthy choices or expedient ones—which often don't end up being very healthy at all. But eating well is often challenging: it's simpler (and cheaper) to breeze into the drive-through than to prepare home-cooked, nutritious meals. Many of my patients struggle with eating healthfully on a fixed income, shopping wisely and just finding the time to make sound choices. To make matters even more complicated, so many of us have varied dietary needs and restrictions based on our GI issues. How do you pick the right diet? This chapter will provide information that will help you make the right choices.

What are the fundamental components of a healthy diet?

A healthy diet is one with the right balance of carbohydrates, proteins and fats. It also includes adequate vitamins and minerals. Calories should be adequate, but not excessive, to support a healthy weight. A healthy diet should help minimize the risk of cardiovascular disease, diabetes or possibly even cancer. And, of course, a healthy diet should be varied and tasty!

The recommended ratios for protein, carbohydrate (sugars and starches) and fat vary. Protein should make up 10 to 35 percent of your total calories, carbohydrates should comprise 45 to 65 percent and fat should be 20 to 35 percent.

The average person in the United States ingests about 15 percent of her calories as protein. A general rule in thinking about the amount of protein you should ingest is that an average woman would need about forty-six grams of protein per day from all sources (rice, beans, legumes, meat, fish, poultry, milk, cheese, eggs). If you're an athlete or pregnant, then you require more protein than an average woman. Some ingested proteins contain all the important building blocks (essential amino acids) to make proteins in your body (complete proteins), and others are incomplete. These incomplete proteins require a complementary source of protein to give you a healthy diet. Animal-based foods are complete protein foods and are called high-quality proteins.

Some sugars and carbohydrates will raise your blood sugar and your insulin level more quickly than others. Foods that cause a rapid spike in blood glucose are classified as having a high glycemic index, and those that cause a gentle spike in blood glucose due to slow metabolism are classified as having a low glycemic index. The glycemic load is a function of both the volume of carbohydrates in the food and the glycemic index. It's glycemic load that affects the blood sugar. In other words, the load looks at how fast a standard portion of a certain food raises blood sugar. For example, watermelon is a high glycemic index food, but because it's mostly water, it is a low glycemic load food. Examples of high glycemic load foods are potatoes, refined breakfast cereals, candy, couscous, white rice and white-flour pasta. Examples of low glycemic load foods are high-fiber fruits and vegetables, bran cereals, and many beans and legumes. Examples of medium glycemic load foods are whole-grain foods, brown rice and oatmeal.

There is no absolute contraindication to eating high glycemic index foods. In fact, varying your diet to include *some* of these foods is okay. The recommendations suggest that one half or less of all the carbohydrates you ingest every day should be from high glycemic index foods. (If you have diabetes, discuss with your doctor how you should be limiting the carbohydrates in your diet.) However, I would strongly recommend that everyone severely restrict the ingestion of candy, high-sugar cereals and sugary drinks.

Saturated and trans fats are not healthy. Most solid fats contain more saturated fats, trans fats and cholesterol, which may raise LDL (low-density lipoprotein), the bad cholesterol. These are from animal fats, hydrogenated vegetable fats, coconut oil and palm kernel oil. In contrast, monounsaturated and polyunsaturated fats are healthy. Most vegetable oils do not contain cholesterol and are low in saturated fats and high in monounsaturated or polyunsaturated fats. Monounsaturated fats are high in nuts (almonds, pecans and hazelnuts), seeds, and canola, peanut and olive oils. Polyunsaturated fats are high in walnuts, corn oil, soybean oil, sunflower seeds and oil, flaxseed and fish oils. New types of solid fats that incorporate plant sterols and products with these fats have been and are being developed (Smart Balance, Earth Balance). They increase HDL (high-density lipoprotein, the healthy cholesterol), decrease LDL (the bad cholesterol) and contain no trans fats.

I'm confused by the various types of food pyramids. Which one can I rely on?

There are several guides for a healthy diet: the U.S. Department of Agriculture's (USDA) MyPyramid; the National Heart, Lung, and Blood Institute's Dietary Approaches to Stop Hypertension eating plan (DASH); and the Harvard School of Public Health's Healthy Eating Pyramid are three of the most highly respected ones. All three of these diets are easily accessed on the Internet and provide a good framework for a nutritious diet. The goal for each diet is slightly different, but the recommendations are similar. Interestingly, they're not part of most Americans' routine diet.

The USDA's guide endorses different types of foods that provide the necessary nutrients recommended by the Institute of Medicine's Dietary Reference Intakes (DRI) and based on the *Dietary Guidelines for Americans* for both sexes and across a range of ages and activity levels. The *Dietary Guidelines for Americans* is updated every five years and will be reissued in 2010. The DASH diet was developed to test whether a diet could reduce hypertension. The DASH eating plan is based on a tested diet of the same name, modified to include more liberal recommendations but similar foods to reduce hypertension and to be heart healthy. The diet is *not* just for those with hypertension— it's a good, nutritious diet for all. The third guide, Harvard's Healthy Eating Pyramid, based on extensive large-population studies that identified nutrients that were associated with reduced rates of chronic disease, emphasizes how often different foods should be ingested in order to reduce the risk of chronic disease. Both the USDA's and Harvard's guides are in the form of pyramids for easy viewing.

Basically, all three guides give the same overall recommendations, but the details vary:

1. Eat more fruits, vegetables, legumes and whole grains.
2. Eat less added sugar and solid fat. All meat should be lean.
3. Eat more plant oils that are high in unsaturated fat. These are cooking oils from plants (canola, corn, soy, peanuts, olive) and nuts that contain healthy oils.
4. Daily exercise is important and necessary. To maintain a healthy weight, the calories that you consume in food have to be offset by the calories that your body burns. Obviously, more calories are burned if you are doing vigorous exercise than if you're sitting around watching TV or a DVD. Thirty minutes or more of moderate exercise five days per week is all most people need to maintain a healthy weight.

The DASH eating plan and Harvard's Healthy Eating Pyramid are very similar to the Mediterranean-style diet, which I endorse. The Mediterranean diet calls for eating a generous amount of fruits and vegetables; eating small portions of nuts; eating healthy fats, such as olive oil and canola oil; eating fish or shellfish at least twice a week; consuming very little red meat; using spices instead of salt to flavor food; and getting exercise, as recommended in all the guides. Moderate ingestion of red wine by some people, as recommended by the Healthy Eating Pyramid, is also incorporated into the diet.

How many calories should I consume per day?

The number of calories a healthy woman should consume is based not only on her size but also on her activity level and age. If you are five feet tall, you will require many fewer calories than your six-foot-tall pal. Women generally need fewer calories after age fifty. The average number of calories that a woman should eat varies by age:

19–30: 2,000–2,400

31–50: 1,800–2,200

51 and over: 1,600–2,200

Below is a basic summary of the recommendations from all three guides for a two-thousand-calorie diet, with the guide differences noted. More details can be obtained online about each guide. (See Appendix 3 for resources.)

Vegetables: 2½ cups per day. Dark green, orange, starchy and other vegetables should be eaten to ensure adequate vitamin ingestion. (Legumes are included in this category for MyPyramid only.)

Fruit: 2 cups (MyPyramid), 2½ cups (DASH), 3½ cups (Harvard's Healthy Eating Pyramid) per day.

Grains: 6–8 ounces per day. (In general 1 ounce is equivalent to 1 slice of bread, ½ cup of cooked rice or pasta, or 1 cup of ready-to-eat cereal.) At least half of the grains should be whole grains (MyPyramid, DASH). Harvard's Healthy Eating Pyramid recommends that less than one-third of the grains have a high glycemic index (including potatoes), which makes most of the grains whole, which is a recommendation of more whole grain ingestion than recommended by the other two eating guides.

Milk products: 2–3 cups of low-fat or nonfat milk per day (MyPyramid, DASH). Lactaid milk or, if necessary, calcium and vitamin D supplements, could be ingested instead. Harvard's Healthy Eating Pyramid recommends mostly supplements (¼ cup of milk only) of calcium and vitamin D. Milk products are a major source of calcium and milk is fortified with vitamin D. You can also get calcium from dark green, leafy vegetables, such as kale and collard greens, and from dried beans and legumes. Soy milk and certain types of orange juice are fortified with calcium.

Meat and beans: 5½ ounces per day of lean meats and beans with 3 cups per week of legumes (MyPyramid); less than or equal to 6 ounces per day of lean meats and 1¾ ounces of nuts, seeds and dry beans (DASH); 4¼ ounces of fish, poultry or eggs (0–2 times per day), ¾ ounce per day of red meat, and 2¾ ounces of nuts, seeds and dry beans 1–3 times per day (Harvard's Healthy Eating Pyramid). This group supplies protein, B vitamins (niacin, thiamin, riboflavin and vitamin B_6), vitamin E, iron, zinc and magnesium in the diet.

Oils: All the diets recommend healthy fats, but the recommendations vary greatly: 6 teaspoons per day (MyPyramid), 2–3 teaspoons per day (DASH), 11½ teaspoons per day (Harvard's Healthy Eating Pyramid).

Other: Saturated fat should be less than 10 percent of all fat ingested. Moderate amounts of alcohol may be healthy in some people (Harvard's Healthy Eating Pyramid): one drink per day for women and two drinks per day for men. (One drink equals 12 ounces of beer, 5 ounces of wine or 1½ ounces of hard liquor.) A recommended healthy fiber intake is about 14 grams per 1,000 calories. The cholesterol intake should be less than 300 mg per day. (The diets ranged from 230 to 271 mg per day.) Sodium intake should be less than 2300 mg per day and potassium should be less than 4700 mg per day.

HEALTHY AND LESS HEALTHY FOOD CHOICES		
Healthy		**Less Healthy**
Protein		
Eggs		Whole milk
Poultry		Cream
Fish		Red meat
Nonfat milk		
Nuts		
Beans		
Legumes		
Seeds		
Soy		
Fat		
Polyunsaturated Monounsaturated		**Saturated Trans**
Fish oil, vegetable oils		Butter
Peanut oil, olive oil, canola oil		Palm kernel oil
Corn oil, sunflower oil		Coconut oil
Trans-free margarine, nuts, vegetable oil		Solid vegetable or meat fat
Oil from nuts, flaxseed oil		
Carbohydrate		
Low Glycemic Index	**Medium Glycemic Index**	**High Glycemic Index**
Pinto, black, kidney beans	Arrowroot, millet, bulgur	Sugary drinks, candy

Lentils, chickpeas	Pearled barley, oatmeal	Refined breakfast cereal
Beans, legumes	Corn, brown rice	couscous, potatoes
Bran cereals	Whole-grain foods	White-flour pasta, white rice
High-fiber fruits and vegetables	No-sugar-added fruit juice	

Are these diets actually practical?

Although a large percentage of Americans know they don't eat well, most of us who think we do eat well don't. Only about 3 to 4 percent of Americans follow all the recommendations for a healthy diet in the *Dietary Guidelines.* The average American eats just three servings of fruits and vegetables a day. This falls short of the five to thirteen servings of fruits and vegetables a day (2½ to 6½ cups per day, depending on one's caloric intake) recommended by the latest *Dietary Guidelines.*

The USDA developed a Healthy Eating Index score, using nutrition recommendations, to assess a person's diet. By this tool, only 17 percent of Americans over the age of sixty consume a "good" diet. This varies by sex, ethnic group, education, weight and age. Women eat more fruit and less cholesterol and sodium than men, but they eat less dairy and meat than men. Women and men with more education have better diets in general, but as people age, healthy nutritional choices decline.

What's the scoop on all these pills I've heard you should take? Do I really need to take a multivitamin, calcium, vitamin D, and omega-3 fish oil capsules?

The answer to these questions depends on whether you obtain all the recommended daily allowances (RDA) of vitamins and minerals in your diet. Most people don't. If your diet meets all the vegetable recommendations for any of the three diet guides above, you should satisfy the vitamin A requirement. (Vitamin A is contained in all those orange and red fruits and vegetables, such as carrots, sweet potatoes, tomatoes, red sweet peppers, mangoes, cantaloupes, apricots and red or pink grapefruits. Green leafy vegetables, such as spinach, collards, turnip greens, kale, beet and mustard greens, green leaf lettuce and romaine, also have vitamin A.)

The vitamin E RDA is not usually met by most people. There are new products that contain vitamin E, and supplements are readily available. When either the MyPyramid or DASH diet is followed, the recommended vitamin E intake usually falls short of the RDA, but if you follow the Healthy Eating Pyramid, the RDA for your vitamin E intake is close to the recommended level.

Calcium is vital to our bone health. Without enough calcium in your bones, you can develop osteoporosis. If you don't get enough calcium in your food (1000 mg if you're aged twelve to forty-nine and 1200 mg if you're fifty and over), you need a supplement. One cup of fat-free milk has about 300 mg of calcium. Other dairy products also contain calcium. Nondairy sources of calcium exist but are not common food choices.

Vitamin D is another nutrient that usually needs supplementing. If you are a sun worshipper, do not wear sunscreen and live in a sunny clime all year round, then you might be an exception. (I don't advocate soaking up the sun and not wearing sunscreen, however.) Vitamin D is made in the skin by the action of sunlight. It's needed for healthy bones, as vitamin D is required for adequate absorption of calcium. Vitamin D may be important for immune health, too. The recommended amount of vitamin D continues to change. For women who do not get sun, the recommended daily intake of vitamin D is one thousand international units (IU) per day. In women who do not absorb the vitamin D well, the elderly and dark-skinned individuals, the requirement for vitamin D is even higher. Adequate vitamin D from all sources for most women should be one thousand IU per day. Vitamin D is available over the counter and varies from four hundred IU to ten thousand IU per tablet or capsule. It is contained in multivitamins and in many calcium tablets in varying doses.

Based upon these findings, I recommend taking a multivitamin and a supplement with calcium and vitamin D. Multivitamins usually contain enough vitamin E and B vitamins, but special groups of people may need to further supplement with other vitamins. What calcium to choose is often an individual choice. The choices are calcium carbonate, calcium citrate, calcium gluconate and calcium lactate. Calcium supplements may contain vitamin K, magnesium or other substances. The amount of calcium varies in each tablet and should be listed on the bottle as the amount per pill or dose. The pills also vary in size and form; some are chewable.

What is more important than the *type* of calcium is that you *actually ingest* the calcium. Don't just stare at the bottle and pat yourself on the back for purchasing it! Calcium is absorbed into the body better if taken with meals. This is especially true if you take acid reducers. Calcium citrate is better absorbed in a nonacid environment than the other types of calcium and would be my recommended first choice for a calcium supplement if you are on this type of medication. Certain antacids, like Tums, contain calcium (calcium carbonate) and should be considered when adding up your ingested calcium. Note: If you have had kidney stones, be sure to consult your doctor before taking extra calcium.

Omega-3 fatty acids have been in the news a lot lately, for good reason. This substance is essential for healthy cell membranes and normal clotting function. It protects against heart disease and maybe stroke, and it may decrease inflammation in such conditions as Crohn's disease. Omega-3 fatty acids are in fatty fish and in some vegetable oils, such as soybean, canola and flaxseed, and in walnuts. They also occur in some green vegetables, such as brussels sprouts, kale, spinach and salad greens. If you don't eat fish, try adding a food with a vegetable source of omega-3 fatty acids. Supplementary omega-3 oil may be beneficial, especially if you have heart disease. Omega-3 fatty acids taken as a supplement or ingested as fatty fish help in the primary prevention of heart disease, result in improved outcomes in patients after a heart attack or with heart failure and aid in lowering plasma triglyceride levels. The American Heart Association supports the use of 1000 mg per day of fatty fish or fish oil supplements in patients with heart disease and 500 mg per day of a supplement or two oily fish meals per week in those without heart disease.

I don't have constipation, but my doctor told me that I should be eating more fiber to try to reduce my bad cholesterol and slightly raised blood sugar. Can this help?

As we discussed in Chapter 5, fiber is material contained in plants that can't be digested by the small intestines. *Soluble* fiber (fiber that can be dissolved in water) passes through the small intestines relatively unchanged until it encounters the bacteria in the colon. The bacteria ferment the fiber, producing gases and short-chain fatty acids, which stimulate the bowels and hold on to water, bulking up the stool. The soluble fiber is concentrated in legumes (dried beans and peas), oats, barley and soy. *Insoluble* fiber passes through the colon relatively unchanged. Insoluble fiber is concentrated in wheat bran, most vegetables and fruit roughage.

There have been many studies looking at the benefits of fiber—and it's not just for constipation control. Overall, those people who have high intakes of dietary fiber appear to be at lower risk for heart disease, stroke, high blood pressure, diabetes and obesity. Increasing fiber intake can lower blood pressure and the bad cholesterol (LDL). Blood sugar can be lowered and the sensitivity of insulin improved in prediabetics and diabetics who supplement with fiber. Furthermore, studies have shown that supplementation with guar, pectin, psyllium or oat β-glucan can reduce LDL cholesterol by as much as 10 percent. You can get these supplements at your local pharmacy.

I skip breakfast most days because I want to sleep! Are there any benefits to breakfast that would make getting up early worthwhile?

You have lots of company, unfortunately. Breakfast skipping occurs in 20 to 40 percent of women on any given day. About 27 percent of teenage girls skip breakfast daily, and 56 percent skip breakfast intermittently. As children age, they have an increasing tendency to skip breakfast. However, this isn't healthy. It has been shown that children and adolescents who eat breakfast are more likely to be thinner than those who skip breakfast. Breakfast consumption may improve memory and test scores in children and adolescents, too. In a five-year study that examined the breakfast-eating habits and their association with body mass index (BMI) of middle school and high school boys and girls, eating breakfast corresponded to a lower BMI. Other studies have shown that those who tend to eat breakfast consume less overall fat and, importantly, less saturated fat than people who skip breakfast. The ingestion of fiber-rich foods or protein, like eggs, early in the day might make you feel full faster and even decrease your desire for a big lunch.

Adults who want to lose weight are likely to see results if they eat breakfast. One study showed major body weight loss and greater energy intake in people who ate breakfast and lunch. Blood lipid and insulin levels also benefit. Bottom line? Set your alarm and eat your breakfast!

I always hear that people should drink eight glasses of water per day. Is this a myth? How much water should I drink to be healthy?

You need to ingest ninety-one fluid ounces (2.7 liters) each day. That's about nine eight-ounce glasses of liquid, with the rest (seventeen ounces) coming from your food. You can even include coffee or tea in your repertoire. Contrary to popular belief, studies show that they do not cause dehydration.

We all think we know obesity when we see it. But technically, what is obesity?

In order to determine if someone has a healthy weight or is overweight or obese, we determine his or her BMI. This doesn't actually measure the fat, but it makes a general estimation by determining a weight-to-height ratio. The BMI is weight in kilograms (2.2 pounds) divided by height in meters (3.28 feet = 1 meter) squared. A BMI of 18.5 to 24.9 is the healthy weight range, a BMI of 25 to 29.9 is the overweight range and a BMI of 30 and above is in the obese range. To see where you stack up, determine your personal BMI at http://www.nhlbisupport.com/bmi.

I've gained fifty pounds over the last two years due to stress, and I see no end in sight. I have to lose weight as I am now overweight, on the border of being obese, and just developed high blood sugar. There are so many diets and weight plans out there. Which one is best?

Dieting can be tricky: after six months of dieting, people tend to start putting weight back on. You need to find something that works for your lifestyle.

So, how do you determine your ideal diet? It's tough. Below is a list of some of the current popular diets and plans. Remember, normal recommended energy intake is: protein, 10 to 35 percent of total calories; carbohydrates, 45 to 65 percent; and fat, 20 to 35 percent.

High-protein diet

This is often called the Zone diet. It is 40 percent carbohydrate, 30 percent protein and 30 percent fat. This diet is a very hard one to follow. Weight loss on this diet is associated with a decrease in the ratio of bad cholesterol (LDL) to good cholesterol (HDL), a good thing. However, long-term studies recently completed do not show any benefit of this diet over other conventional diet plans. Claims that it should improve immunological health by modulating anti-inflammatory substances called eicosanoids remain to be proven.

Low-carbohydrate diet

Atkins

Atkins allows twenty grams per day or less of carbohydrates in the initial phase of the diet and fifty grams per day or less of carbohydrates for ongoing weight loss. By restricting carbohydrate intake, the body burns its own fat. This raises the level of a substance called ketones (breakdown products of fat), which make you feel less hungry in the early phase of the diet, when there are severe restrictions on fruit, vegetables and fiber. In the first stage you are allowed three cups of loosely packed salad or two cups of salad with two-thirds cup of certain types of cooked vegetables each day. The twenty grams of carbohydrate is calculated by subtracting the amount of fiber present from the total amount of carbohydrate. In the second stage carbohydrates are slowly added back until weight loss stops. Nuts and seeds are allowed, but only a small selection of fruits. (Blueberries, raspberries, strawberries, cantaloupe and honeydew are okay.) Milk, white flour or products from white flour (like pasta), and white potatoes are forbidden.

This is a controversial diet. The concerns are that the low fruit, vegetable and fiber intake with high fat may not be good for the heart in the long run. Calcium loss by the kidneys can be high, and this type of diet can put stress on the kidneys and liver. However, more long-term studies will be needed to determine whether the current concern over the diet will be verified or dismissed.

South Beach Diet

This wildly popular diet is a three-phase eating plan. It's high in protein, low in saturated fat and low in carbohydrates with a high glycemic load. During the first two-week stage, fruit, bread, baked goods, pasta, and beer and wine are banned. Lean meats, low glycemic index vegetables, low-fat cheese, nuts and eggs are allowed. Eight to twelve pounds may be lost in the first stage, but most of the weight loss is water weight, stored in the tissues, and not loss from fat weight—your true goal. During the second stage some carbohydrates are allowed. Those that are chosen should have a low glycemic index as discussed above. Your goal should be one to two pounds of weight loss per week after the initial six weeks. The final stage, to maintain weight, is done by limiting portions and continuing to eat low glycemic index foods. Healthy fats are encouraged.

Balanced diet

Weight Watchers

Weight Watchers stresses limited calories and regulated portion sizes. It allows for variety and support from a dieting community and emphasizes a lifestyle of healthy food choices with exercise. Points are assigned to foods, with low points going to high-fiber, low-fat and low-calorie foods. A dieter is allowed a certain number of points per day. You choose your foods, which allows for variety and taste.

Jenny Craig

All foods must be purchased from Jenny Craig. This puts structure on the diet but is limiting for different palates. Portion control is key. Nutritional consultations and coaches are part of the program, and so is exercise.

Low-fat diet

Ornish

This is a high-fiber vegetarian diet developed by Dean Ornish, with no more than 10 percent of calories from fat allowed. Snacking when you desire is encouraged. People feel hungry on this diet and snacking helps relieve the hunger. Getting enough protein requires vigilance.

LEARN

This is a low-fat, moderately high carbohydrate diet. Between 55 and 60 percent carbohydrates is allowed, and fewer than 10 percent of calories come from saturated fat. Calorie restriction, exercise and behavioral modification are important.

How do these diets stack up against one another in the short and long term? Unfortunately, when careful evaluation was done to determine whether an individual actually stuck to the diet, the compliance was very poor. In the short term (the first six months), analysis of the studies conducted between 2003 and 2006 suggests that the low-carbohydrate diets seemed to result in more weight loss than the low-fat diets. However, by twelve months, this was no longer true.

The Atkins, Ornish, Weight Watchers and Zone diets compared equally after twelve months, with a weight loss of 4.8, 6.0, 4.9 and 7.3 pounds, respectively, in those who dieted for twelve months. Unfortunately, the dropout rate was 35 to 50 percent in each group. The Ornish group of dieters ate closer to 17 percent in fat calories, rather than the recommended 10 percent. Each of these diets improved the ratio of good cholesterol to bad cholesterol by 10 percent.

In a separate study of premenopausal, obese, nondiabetic women on the Atkins, Zone, Ornish and LEARN diets, women at twelve months lost more weight on the Atkins diet (10.3 pounds) compared to the women on the other three diets (3.7, 5.7 and 4.8 pounds, respectively). Also, good cholesterol was slightly higher and blood pressure was slightly reduced for women on the Atkins diet.

A third study examined the results at two years of four diets that varied in nutrient composition with high and low fat and average and high protein. The biggest weight loss was in the first six months—a modest 13 pounds (6 kilograms). However, by twelve months the weight loss became weight gain and by twenty-four months on average the total weight loss was only 6 to 9 pounds (3 to 4 kilograms). In a small group of dieters a large amount of weight loss was achieved. The takeaway was basically that people able to stick to a diet lost more weight, and those dieters who lost more weight attended more nutritional counseling sessions.

In the future more specific recommendations for the right type of diet for you may be determined by taking a swab of your cheek and analyzing your genetic makeup. In a preliminary study researchers at Stanford University have been correlating certain genes with weight loss, trying to determine who would benefit most from a low-carbohydrate versus a low-fat diet. This provocative study, utilizing a simple test to predict the best weight-loss diet for a person, will require larger studies for verification.

To lose weight successfully:
- Choose a diet that you think you can follow and stick to it.
- Build a community that will support you. This can be a program like Weight Watchers or just friends and family.
- Make healthy food choices. Eat healthy fats and carbohydrates with a low glycemic index. There are sources online that rate foods for their glycemic index and glycemic loads. The chart on page 162 gives a general idea of the classification of some foods.

- Exercise every day if possible, for at least thirty minutes. Burning calories will help you lose weight, and the exercise will make you feel better overall. Start slowly and build up the amount and difficulty over time.
- Eat slowly to give your body a chance to feel full.
- Discuss your diet goals and choices with your physician or nutritionist. There are many plans available on the Internet. The U.S. government makes helpful suggestions on its nutrition site (http://www.nutrition.gov).

How can I trick myself into feeling full?

This is a popular question! I recommend a few different strategies. Know that proteins always makes you feel fullest, followed by carbohydrates and then fats. A high-protein meal is going to fill you up faster than gorging on bowls of cereal. Furthermore, if you eat slowly, you'll give yourself time for your body to adjust to the food you have eaten and release substances from your intestinal tract that tells your brain that you are full. By stuffing yourself fast, you can "beat" your body's stop signal.

One popular strategy is to drink two glasses of water before mealtime. In one study obese individuals who drank two glasses of water before breakfast consumed fewer calories than those who did not have prebreakfast water.

I have heard that the Specific Carbohydrate Diet may be helpful for my irritable bowel syndrome. What is it and will it help?

The Specific Carbohydrate Diet is a strict grain-free, lactose-free and sucrose-free diet that was designed for people with Crohn's disease, ulcerative colitis, celiac disease, inflammatory bowel disease and irritable bowel syndrome. It was developed by Sydney Valentine Haas, M.D. Elaine Gottschall helped to popularize the diet after using it to help her daughter recover from ulcerative colitis. (Gottschall continued doing research on the diet and later wrote her own book, *Breaking the Vicious Cycle: Intestinal Health Through Diet.*)

Some may consider the diet extreme, but some tenets are worth extracting. The theory is that carbohydrates, being forms of sugar, promote and fuel the growth of bacteria and yeast in the intestines, causing an imbalance and eventual overgrowth of bacteria and yeast. According to Dr. Haas, a number of illnesses could then develop from this altered digestive balance:

- Crohn's disease
- Ulcerative colitis
- Celiac disease
- Inflammatory bowel disease (IBD)
- Irritable bowel syndrome (IBS)
- Chronic diarrhea
- Spastic colon

Dr. Haas designed this diet to correct the imbalance by restricting the carbohydrates available to intestinal bacteria and yeast. Only carbohydrates that he believed to be well-absorbed are consumed on the diet so that intestinal bacteria have nothing to feed on. This, he proposed, would help correct the bacterial overgrowth and related mucus and toxin production.

While there's no hard evidence pointing to its effectiveness, I have found that a number of my patients with Crohn's disease, ulcerative colitis and irritable bowel syndrome have experienced relief with this diet. Here's a general list of foods to avoid and foods to seek out.

Avoid:
- Canned veggies
- Canned fruits, unless they are packed in their own juices
- All cereal grains, including flour
- Potatoes, yams, parsnips, chickpeas, bean sprouts, soybeans, fava beans, seaweed
- Processed meats, breaded or canned fish, processed cheeses, smoked or canned meat
- Milk or dried milk solids
- Buttermilk or acidophilus milk, commercially prepared yogurt and sour cream, soy milk, instant tea or coffee, coffee substitutes, beer
- Cornstarch, arrowroot or other starches; chocolate or carob; bouillon cubes or instant soup bases; all products made with refined sugar, agar, carrageenan or pectin; ketchup; ice cream; molasses; corn or maple syrup; flours made from legumes; baking powder; medication containing sugar; all seeds

Enjoy:
- Fresh and frozen veggies and legumes
- Fresh, raw or dried fruits

- Fresh or frozen meats, poultry, fish, eggs
- Natural cheeses, homemade yogurt, dry curd cottage cheese

The Specific Carbohydrate Diet does contain several foods that can be problematic for some people. Those individuals with fructose malabsorption or lactose intolerance will still have a source of those foods that could cause symptoms. In those with fructose malabsorption and IBS, dietary restrictions of fructose and fructans have been shown to help symptoms. An elimination diet is often the best way to find a diet for a particular person with IBS.

I have celiac disease, but I am still unsure of what I should be eating and avoiding. Can you give me more guidance?

It is not always intuitive that a food is safe to eat. Avoid wheat (farina, spelt, semolina, durum, graham flour, bulgur, Kamut, matzo meal), barley, rye and triticale. The following food products are safe: corn, rice, potatoes, soybeans, tapioca, buckwheat, millet, amaranth, quinoa, arrowroot and carob. Gluten-free bread and other baked products are available in many markets. Meats, fish, fruits and vegetables are safe foods. Distilled white vinegar is safe. (Malt vinegar is *not* safe.) Gluten-free soy sauces are available.

Food labels are helpful in sorting out whether gluten is an included ingredient. However, when you are eating out and not seeing the labels, it can be a problem to figure out if a food is safe. And if gluten-free is not designated, then there is no guarantee of safety. Additives can be a problem. Vegetable or plant protein should come from soy or corn. Malt or malt flavoring should be derived from corn. (This leaves out ordinary beer! However, good-flavored gluten-free beer is available.) Modified starch or modified food starch is only safe if it comes from arrowroot, corn, potato, tapioca, waxy maize or maize. Likewise vegetable gums have to be from safe foods: carob bean (locust bean), cellulose, guar, gum arabic (acacia), tragacanth, xanthan and vegetable starch. Flour or cereal products need to be made with pure rice flour, corn flour, potato flour or soy flour. If you can make all your own bread, know that your food is not cooked with food "contaminated" by wheat, barley or rye, and stay away from additives—no problem! However, this can be quite a challenge.

There are specific cookbooks and many online recipes for people with celiac disease. When trying to adjust to a gluten-free diet, it is easier to start with recipes that are gluten-free and then you can modify recipes not written for a person eating gluten-free. There is no problem with meats, legumes, fruits and vegetables. As for grains, it is helpful to familiarize yourself with how you substitute different grains for banned substances. Finally, note that lipsticks, lip balm, Play-Doh and postage stamps often have gluten in them. Also, look at all the inactive ingredients in your medications and supplements to make sure there is no gluten.

I want to eat healthy foods, but my budget is limited. It's just cheaper to buy potatoes or pasta, or to grab something on my way home from work. Do you have suggestions?

Unfortunately, this is a common problem. It requires creative thinking and a lot of attention to local information on sales and specials. I think it's important to buy in bulk. This might not seem practical if you have no storage space. However, if you get together with a few friends, you could share the items and split the cost. Large supermarkets or food-club stores tend to be cheaper than organic food stores. Prices also vary according to location. If you have the luxury to shop at more than one store, compare prices. Try to buy fruits and vegetables when in season. Branch out. Don't always select the same fruit or vegetable. Try cheaper vegetables when available. If fresh vegetables are too expensive, compare prices for the frozen and canned vegetables. (Remember when buying frozen or canned products to carefully inspect the label for additives. You should try to avoid a lot of added salt or other substances.)

Fish is very healthy, but I realize it's often very expensive. Look for specials! Try different types of fish. Chicken and turkey are usually cheaper alternatives to fish and are also good for you. If you buy in bulk (over three pounds of chicken), the price per pound usually drops. If you make a big turkey or chicken dish, you can freeze the leftovers. But be sure to put on a label with the date you made the dish before putting it in the freezer, so that you can try to eat it within four months for a better flavor.

Soups are very economical. You can use cheaper parts of a chicken or the bones from a chicken or turkey. Throw in some vegetables—carrots, celery, peas—beans, pasta, rice, bulgur or noodles and you have several delicious meals. Experiment with spices. Spices can make potentially boring meals more exciting. Dried spices will last a long time although they may lose some potency after a while.

Ground beef is usually inexpensive. Just make sure you buy very lean beef or sirloin ground beef if possible to make sure that there is less fat. You may be able to substitute ground turkey for about the equivalent price. The ground turkey or beef could be made into meat loaf or chili for a nutritious meal.

Finally, experiment with dried lentils and beans. These are usually inexpensive. They can be reconstituted by soaking them overnight and then can be used as the principle protein source in combination with vegetables and, if desired, a small amount of meat.

WHAT YOU NEED TO KNOW:

1. A healthy diet provides the right balance of carbohydrates, proteins and fats.

2. Sugars with a high glycemic index, such as sugary drinks and candy, can cause a rapid rise in blood glucose. However, it is the glycemic load (volume of carbohydrates and glycemic index) that determines how fast the blood sugar will rise.

3. Saturated and trans fat are not healthy.

4. Monounsaturated, polyunsaturated and omega-3 fats are healthy fats.

5. Adequate ingestion of vitamin D, calcium, vitamin E and omega-3 fat often requires supplementation.

6. Fiber can improve constipation while lowering blood pressure, bad cholesterol (LDL) and blood sugar.

7. The best diet for you is the one that you can follow conscientiously over a long period of time.

8. Exercise is essential for overall health and achieving and maintaining a healthy weight.

Chapter 10
Doctors' Visits and Medications

"Never go to a doctor whose office plants have died."
—Erma Bombeck

There's nothing more unsettling than leaving a doctor's appointment with more questions than answers. This chapter is a user-friendly guide to navigating the medical maze. How can you get the most out of your GI visit? How do you find a GI doctor you connect with? This chapter outlines the basics.

What should I look for in a physician, and how do I find a good one?

Most of all, find someone who listens and who doesn't make you feel rushed. Also, look for someone with a strong network of contacts, should you need a referral; someone who communicates results and someone with whom you feel comfortable; and, most of all, someone who treats you with respect. It's important for you to have a primary care physician (PCP), who can coordinate all your care. Your PCP is an excellent source for a referral. Other sources are your friends, relatives and often colleagues. You can find information about a physician's interests and credentials online. However, credentials don't necessarily tell you about bedside manner or her approach to a problem. If you have a problem that might at some time require hospitalization, you want to pick a physician who can admit you to your chosen hospital. At an academic medical center or community hospital affiliated with a medical school, you're likely to have a physician who is up-to-date, teaching and often doing research in subspecialty areas in gastroenterology. You will also likely have residents and fellows who are in training affiliated with your care. This may or may not be to your liking. There are many excellent community-based, up-to-date and creative gastroenterologists. Your specialists, including your gastroenterologist, should communicate their findings to your PCP.

I'm going to the GI doctor for the very first time. What information should I bring? What should I ask?

1. Come with an outline of your medical history. What is "normal" for you?

2. When did things change?

3. How did things change?

4. What are your specific intestinal symptoms? (Bleeding, heartburn, etc.)

5. Are there any other problems? (Chest pain? Fever? Rashes?)

6. What's your diet? It's often helpful to jot down a typical day's diet. If your diet varies, write down the foods you eat over the course of a couple of days. This is particularly important if you're gaining or losing weight or if food precipitates your symptoms.

7. Bring a list of your past operations and hospitalizations.

8. Your doctor will ask you about your family's medical history. If you do not have it (and you're on good speaking terms with your relatives!), you might ask them if they have any medical illnesses. Those in which your gastroenterologist might be particularly interested include Crohn's disease, ulcerative colitis, colorectal cancer and other cancers, autoimmune diseases like rheumatoid arthritis and systemic lupus erythematosus, and celiac disease.

9. Ask how your test results will be communicated to you and your primary care physician.

10. Determine the best way for you to reach your GI doctor—by phone, by email, or on a patient website.

11. If you need an endoscopy or colonoscopy, ask if he or she will perform it. Also find out what type of anesthesia will be used: conscious sedation (a sedative

and a pain medication given under the direction of the gastroenterologist); propofol anesthesia (heavily sedates you or puts you to sleep), under the direction of the gastroenterologist or by an anesthesiologist; or general anesthesia, in which a tube for breathing is inserted into your trachea (rarely used). If you have a preference or if you would prefer no anesthesia, be sure to discuss this ahead of time with your gastroenterologist.

12. Find out how often you need to return to the office.

My doctor often prescribes generic medications. Is generic as good as name brand?

Usually, yes. The FDA approves only generic drugs that deliver into the bloodstream 80 to 125 percent of their active ingredient relative to name-brand counterparts. In most people with most drugs, this variation is not a factor. However, in an extremely sensitive patient, this variation in amount may change the effectiveness of a medication or its side effects. In addition, some generic medications contain lactose, which could result in abdominal pain for a lactose-intolerant person. A medication that remains in the gut and acts locally might not dissolve as well or be dispersed for optimal effectiveness in the intestines. (One example of a medication that works locally in the large intestines in which some generic preparations have been reported to not work as well as the name-brand medication is cholestyramine, the generic for Questran. My patients have reported a decreased effectiveness when they have noticed the cholestyramine forming globules rather than evenly dispersing.) Anytime a generic is changed from one manufacturer to another, there's the possibility that a dye or other compound could be added. I have some patients who appear to have side effects from a specific generic medication but not from another manufacturer's generic or name-brand formulation. If you try a generic and you don't tolerate it, or you find it less effective, then you might have to use a name-brand medication instead.

This is sometimes problematic because insurance companies often don't want to cover name-brand medications if a generic medication is available. Your doctor will most likely have to obtain a pre-authorization for the drug, and even then there's likely to be a more expensive co-pay than if you were able to take the generic.

If I have chronic stomach problems, which pharmacy staples should be in my medicine cabinet?

Pepto-Bismol! That's especially helpful for food poisoning or traveler's diarrhea. Imodium (loperamide) may help also for diarrhea. An antacid or H2 blocker, like Tums, Maalox or Mylanta, Zantac or Pepcid, might help heartburn. And for constipation, fiber, fluids and those delicious prunes are the best remedies.

Which over-the-counter items aren't all they're cracked up to be?

- **Stool softeners:** Studies show they're not particularly effective.
- **Laxatives:** Talk to a physician before using these frequently. The magnesium preparations are gentle but should not be used if you have kidney problems. Many laxatives are effective, but several of them can cause problems. Recently a colonoscopy preparation, Fleet Phospho-soda, sold over the counter as a laxative, was voluntarily removed from the market by the manufacturer because of an FDA warning that it could cause kidney damage in some people. Rarely in some people bisacodyl has caused abdominal pain, ischemic colitis and bleeding. Therefore, be prudent when using laxatives. For any questions, consult your physician.
- **Imodium:** It decreases diarrhea but might not reduce cramping, and if you have a potential bacterial infection or IBD, you should consult a doctor before treating yourself.

I often see drug advertisements on television. Are these highly advertised medications better than older medications or the ones that I am currently taking?

Remember that the advertisements are a result of the pharmaceutical industry trying to increase market share for a drug. In other words, the company wants to make money. The industry is depending on you to get the doctor to do what is in the industry's best interest—to buy their product. Having said that, it's possible that heavily advertised drugs could have advantages; just because they're on TV doesn't make them bad. However, if you are on a generic medication for your condition, it's almost certain that the advertised drug will be more expensive. It might not be worth taking a new medication with only a marginally better effect if it has a much higher cost.

It's recently become much more difficult for pharmaceutical representatives to have access to your physician. Rightly so. The practice of influencing your physician to use a particular product by bestowing gifts or perks should be eliminated. However, now that this has happened, the pharmaceutical companies have to find different ways to get their messages to your doctor. Unless a control is put in place on direct-to-consumer advertising on television, radio, billboards, Internet sites and in mailings, it's likely you will see more, not less, of this sort of advertising activity. Basically, if you're doing well on a medication, don't ask to change to another one because you saw a persuasive advertisement.

What are the hallmarks of a good doctor-patient relationship?

Much of this is very individual. However, there are themes that are common for all of us. When I asked this question of my patients, Cynthia provided the following:

What I Look For in a Doctor:

Compassion/Understanding

I want my doctor to realize I'm coming to see her because I have an ailment, am not feeling well and need some sort of explanation of what's wrong.

Relating the Problem

I need my doctor to explain to me (not in medical jargon, but in layman's terms) what's going on, so that I comprehend the problem, diagnosis and procedure.

Guidance

I look to my doctor for guidance and direction. I'm placing my life in her hands and trusting her with it.

Trust

You have to have a doctor to whom you can relate and in whom you trust; to whom you are more than a hospital number; and who will be honest with you and not disguise the problem.

Relationship

You have to pick a doctor with whom you feel comfortable so that you can come to her and be personal and honest and really tell her what's wrong, because if you don't, she can't help you.

Level of Comfort/Relaxation

If you have a good relationship with your doctor, no matter what the problem, you should leave the office feeling more at ease than when you walked in.

Reassurance

I want my doctor to reassure me that she is in my corner and looking out for my best interests regarding my health.

Don't Settle

Do not just settle on a doctor. Finding the right doctor makes all the difference in how you feel and how you feel you'll be cared for. Your relationship with your doctor is important. You want a doctor who, although you may be sick, you actually look forward to seeing because of how she treats you, how she diagnoses you and how you feel leaving her office. The right doctor—one whom you trust and feel comfortable with—is the key! It makes all the difference in your care and is a lift to your spirits.

Coping with a chronic condition can be challenging and often embarrassing. How can I make it to my child's play, to my friend's wedding, to work on time?

Below are helpful hints from another one of my patients.

Tips for Coping with GI Symptoms:

Get Over the Self-Blame.

IBD (and probably IBS) can be so confusing and embarrassing. I felt much better medically once I stopped blaming myself for having the disease. Moving beyond self-blame includes educating others about the disease, mostly so that they stop telling you to "just relax" and "not worry so much"—as if you can entirely control the disease and symptoms.

Get a Fabulous Doctor.

One of the best decisions I've made was to switch physicians. My doctor listens, she's a problem solver and she really cares. She was the first person to ever ask me about IBD and my sex life!

Participate in a Mind/Body Program.

The combination of meditation, yoga, participant support and understanding, and learning mindfulness made a big difference in my life.

Know Your Limits.

Get over denying that you have a medical condition, and realize that you have limits. Make sure not to push yourself too hard trying to be everything to everybody (we women do that often), and take care of yourself. That means figuring out what your self-care plan is (this is different for every person, I imagine) and then doing it! The self-care plan could include eating well and getting enough rest, but it may also mean seeing friends often, reading good books and giving yourself permission for downtime. (It can be so hard for women to just sit!)

Don't Limit Yourself.

This sounds like a contradiction of "know your limits," but what I mean is, don't let your medical issues isolate you and keep you from doing the things that make you happy. When you are afraid of having to use the bathroom unexpectedly or urgently, it's easy to just stay home and avoid the unexpected. But this means missing out on so much. For me, I often have to remind myself that it's not the worst thing in the world to be incontinent or to use a gas station bathroom. So, I plan for it—I carry extra underwear, pads and baby wipes in my bag at all times. In fact, I wore an adult diaper at Disney World—just so I could enjoy this trip with my family. I actually never had an accident, but knowing I wouldn't worry about the embarrassment meant that I focused only on having fun with my parents, sister, nephew, husband and two amazing sons. When I look back on all our family trips, I never think about having worn Depends or going to the bathroom in the woods (we like to go on hikes)—all I see are the wonderful family memories.

Have a Sense of Humor.

You absolutely have to find ways to laugh about passing gas! Poop jokes are helpful, especially with immediate family members. Humor is the best way for helping my kids deal with the frequent pit stops on our road trips.

A third patient, Beatrice, who suffered with rectal pain for a long time before a successful treatment was found, has other useful tips and hints.

1. Find a surgeon/doctor/physical therapist who will listen to your complaints and calls back in a timely fashion when needed. Be sure you have chemistry with this person.

2. Change doctors when you feel that you are not getting any creative solutions to your problem.

3. Have a good PCP who can coordinate with your specialists and be available when needed.

4. You have a right to feel sorry for yourself! Cry when you must, within moderation. It lets out stress. Find a good friend who doesn't mind listening to you and sympathizes, as this takes stress off your partner and family. There are understanding friends out there!

5. Rely on your family for help to and from doctors' appointments, help with shopping, laundry, etc. But try to do one little thing a day, to feel that you've done something productive, even if it's just putting in a load of laundry.

6. Be aware that your partner and family, if you have them, are suffering with you and feel helpless not being able to help you, which is why a good doctor is so important to restore hope in your life.

7. Try not to give up seeking solutions. Go on the Internet for information, keep questioning your doctors and ask them for more ideas. Try anything that sounds like it might work (as long as it is not potentially harmful).

WHAT YOU NEED TO KNOW:

1. Prepare for your visit to your gastroenterologist by writing down your symptom history, diet and the medications that you have tried.

2. Find a physician with whom you can relate and openly communicate your problems.

3. Understand how you will be informed of test results, how the plan for your therapy will be communicated and how the results will be monitored.

4. If you do not improve, you can always seek another opinion.

5. Do not be afraid of taking generic medications. In most cases generic drugs are as good as name-brand drugs. However, if you detect a difference in the effectiveness of your medications with a new prescription, be sure and discuss it with your physician.

GLOSSARY

Ablation: To remove or get rid of a structure.

Actinobacteria: Bacteria that are more abundant in the intestines of obese twins compared to lean twins.

Acupuncture: A technique by which fine needles are inserted and manipulated in the body for therapy or pain relief.

Adenomyosis: A condition in which the uterine-lining cells grow into the muscles of the walls of the uterus.

Adhesions: Scar tissue.

Adrenal glands: The endocrine organ above the kidneys that releases hormones that are important for the stress response.

Anal fissure: A cut in the anus that can cause pain and bleeding.

Anal sphincter: The muscular structure responsible for control of stool evacuation. The external sphincter is under voluntary control, but the internal sphincter is not.

Androgens: Male hormones, like testosterone.

Anemia: A decrease in the normal number of red blood cells.

Angina: Chest pain or discomfort that occurs when your heart muscle doesn't get enough oxygen.

Anorexia: Loss of appetite.

Anorexia nervosa: An eating disorder that involves limiting the amount of food that is ingested.

Antibody: Protein made by your immune cells in the body to fight against bacteria, viruses and other substances.

Antispasmodic: A substance that decreases nonproductive muscle contractions, called spasms.

Appendicitis: Inflammation of the appendix, which is the tubelike structure with a blind end that is attached to the bottom of the colon, to the part called the cecum.

Autoantibody: An antibody made against a substance normally found in your body.

Bacteria: Living single-celled organisms that can be seen under a microscope. Some can cause disease, while others can cohabitate in our bodies without negative consequences, and even with positive benefits.

Bacteroidetes: A type of bacteria in greater abundance in the intestines of lean twins compared to obese twins.

Barium test: An X-ray test in which a suspension of barium sulfate is used as a contrast agent. Barium tests include barium and video swallows, upper GI series, small bowel X-ray, barium enema and defogram. A CAT scan uses dilute barium contrast.

Barrett's esophagus: A condition caused by acid reflux in which there is a change in the lining cells of the esophagus. It can progress to esophageal cancer in some people.

Bifidobacteria: A type of bacteria found in the gut and in probiotics. These are considered healthy bacteria.

Bile: A fluid made in the liver that contains bile acids, cholesterol and phospholipids. It is important for digestion and absorbing fat.

Biofeedback: A technique in which people learn to improve their health through control of different bodily functions. This technique is used to improve anal sphincter tone in patients with leakage of stool and to retrain the pelvic and anal sphincter muscles to function properly to evacuate stool in patients with constipation.

Bisphosphonate: A medication used to treat osteoporosis.

Bulimia: An eating disorder in which there is recurrent binge eating, followed by vomiting and/or other purging.

CA-125: A glycoprotein that is elevated in some tumors, such as ovarian cancer.

Carbohydrates: Sugars (simple carbohydrates) and starches (complex carbohydrates)—one of the three basic food classes.

CAT scan (Computerized Axial Tomography): A test, utilizing X-rays, to visualize the structures in the body. The procedure is done with the patient lying down and surrounded by the scan device.

Cautery: The process of destroying abnormal tissue by burning, searing or scarring.

CBC: Complete blood count. A blood test measuring the white blood cell count, hemoglobin, hematocrit (percentage of red blood cells), red blood cell count, platelet count and characteristics of the red blood cells.

Celiac disease: An autoimmune disease of the small intestine caused by a reaction to gliadin, a portion of the gluten protein found in wheat and many other grains.

Cholecystitis: Inflammation of the gallbladder. The most common cause is gallstones.

Chronic fatigue syndrome: A condition of prolonged fatigue or tiredness.

Clostridia difficile: A bacterium that overgrows often after antibiotics use and that can cause diarrhea or colitis.

Colitis: A general term for inflammation of the colon.

Colon: The part of the large intestine that ends in the rectum and anus. The parts of the colon are the cecum, ascending colon and hepatic flexure on the right; the transverse colon across the upper abdomen; the splenic flexure, descending colon and sigmoid colon on the left; and the rectum.

Colonic: Referring to the colon. This term is often used synonymously with colon hydrotherapy, a procedure in which large amounts of water are used to irrigate the colon.

Colonoscopy: A procedure in which a long, flexible tube with a light is advanced from the rectum through the colon and often into the last part of the small intestine, the terminal ileum. Polyps can be removed and samples of tissue taken through the scope.

Constipation: Fewer than three bowel movements per week; a decreased frequency from or amount of your normal stools; less than one hundred grams (about a quarter pound) of feces; or difficulty getting the stool out.

Corticotropin-releasing hormone (CRH): A major mediator of the stress response, made in the part of the brain called the hypothalamus.

Cortisol: The "fight or flight" hormone made in the adrenal gland.

Cortisone: A steroid hormone that decreases inflammation.

Crohn's disease: A chronic inflammatory condition that can affect any part of the GI tract, but most commonly the small intestine and colon. It is associated with many other conditions, including inflammation of the anus and the surrounding area, joint pain and inflammation, skin rashes, gallstones and kidney stones. It is one of the main types of inflammatory bowel disease.

Defogram: An X-ray study to examine the process of elimination of waste from the rectum. The evacuation process is recorded following introduction of barium into the rectum and sigmoid colon, and barium paste into the vagina.

Dermatitis herpetiformis: An autoimmune blistering, itchy skin rash associated with celiac disease.

DHEA and DHEA sulfate (dehydroepiandrosterone): The most abundant steroid in the blood, which is produced by the adrenal gland and in the brain.

Diarrhea: More than two stools per day; an increased number of bowel movements per day; an increased stool bulk and amount; or stool that is watery, loose or mushy.

Diverticulitis: Inflammation of a diverticulum.

Diverticulosis: A condition where there is the presence of outpouchings in the colon.

Diverticulum (plural diverticula): An outpouching of the wall of a digestive tract organ: esophagus, stomach, small intestine or colon.

Down syndrome: A chromosomal disorder in which there is an extra chromosome 21, causing multiple physical and developmental abnormalities.

Dyschezia: Painful bowel movements.

Dysmenorrhea: Painful menstruation; menstrual cramps.

Dyspareunia: Painful sex.

Dysuria: Painful or difficult urination.

Endometriosis: A condition in which the cells that line the walls of the uterus take up residence outside the uterus, that is, in the pelvis, on the bowel or elsewhere.

Endoscope: A scope with a flexible tube with a light that is used to look inside a body cavity.

Endoscopy: A general term for the procedure in which a flexible tube with a light source is inserted through the mouth, anus or stoma into the intestinal tract in order to visualize the lining of the digestive tract.

Enema: A procedure in which fluid is inserted into the rectum via an inserted tube.

Enterography: A radiology test to visualize the small intestine.

Epinephrine: A hormone and neurotransmitter that increases the heart rate, contracts blood vessels, dilates the air passages and is important in the stress response.

ERCP (endoscopic retrograde cholangiopancreatography): An upper endoscopic procedure in which a flexible tube with a light source is inserted through the mouth and advanced into the common bile duct and pancreatic duct to evaluate and treat stones, tumors or narrowing in the ducts.

Esophagitis: Inflammation of the esophagus.

Esophagus: The part of the digestive tract between the mouth and stomach through which food passes.

Estrogens: Primary female sex hormones.

FDA (Food and Drug Administration): The U.S. regulatory agency responsible for food and drug safety.

Fecal incontinence: The leakage of stool out of the rectum.

Fiber: A material contained in plants that can't be digested by the small intestine.

Fibromyalgia: A chronic pain condition characterized by tender points in joints, muscles, tendons and soft tissue.

FODMAPs: Fermentable Oligo-, Di- and Mono-saccharides and Polyols. Sugar compounds that can be digested by the gut bacteria to form gas.

Gallbladder: The digestive tract organ connected to the liver by ducts and where bile is stored.

GERD (gastroesophageal reflux disease): The process of any fluid or food coming up the esophagus, not just acid.

GI (gastrointestinal): Refers to the gastrointestinal tract or a physician who specializes in gastroenterology, a gastroenterologist.

Gliadin: The alcohol-soluble portion of gluten, which damages the intestine in people with celiac disease.

Gluten: A nutritional protein stored in wheat to which people with celiac disease react adversely.

Glycemic index: A ranking of foods based on how they affect the blood glucose level.

Glycemic load: A measurement of how fast a standard portion of a certain food raises blood sugar.

Gonadotropin-releasing hormones: Hormones released by the hypothalamus, important for the menstrual cycle.

H2 blocker, H₂RA, H2 receptor antagonist: A medication that blocks the histamine receptor type 2, thereby decreasing acid production in the stomach.

Halitosis: Bad breath.

Hashimoto's thyroiditis: A common autoimmune condition in which there is inflammation of the thyroid gland.

HDL (high-density lipoprotein): A lipoprotein that transports cholesterol in the blood that is considered the good type of cholesterol, associated with lower risk of heart disease.

Heartburn: A burning feeling in the middle of the chest caused by acid.

Helicobacter pylori: Bacteria linked to stomach and duodenal ulcers.

Hemorrhoids: Dilated veins of the rectum that when inflamed can cause bleeding, itching and pain.

Hiatal hernia: The protrusion of part of the stomach into the chest, above the diaphragm.

Homeopathy: A form of alternative medicine that treats people with very diluted substances that are thought to cause similar symptoms to those of the condition being treated.

Hyperemesis gravidarum: Excessive vomiting during pregnancy.

Hypothalamus: A part of the brain that makes hormones that control a number of bodily functions, one of which is the menstrual cycle.

Hypothyroidism: The condition of having decreased function of the thyroid.

Hysterectomy: Surgery to remove the uterus. A total abdominal hysterectomy with salpingo-oophorectomy removes the uterus, fallopian tubes and ovaries.

Ileum: The third portion of the small intestine, which connects to the colon.

Indigestion: A vague feeling of abdominal discomfort that can include heartburn, belching, bloating and nausea.

Inflammatory bowel disease (IBD): Chronic inflammatory disease of the small and large intestines, of which the principal types are ulcerative colitis and Crohn's disease.

Interstitial cystitis: The condition of chronic inflammation of the bladder wall.

Inulin: A carbohydrate that is not digested by the body but is used as a food supply by the bacteria in the gut and considered a prebiotic.

Irritable bowel syndrome (IBS): A chronic condition consisting of abdominal pain or discomfort and a change in stools.

Ischemia: A decrease in blood supply to an organ.

Ischemic colitis: A lack of blood flow to the colon that can cause bleeding.

IVF (in vitro fertilization): A process of manually combining an egg and sperm outside of a woman's body.

Jejunum: Second portion of the small intestine, between the duodenum and ileum.

Lactase: The enzyme that digests the sugar lactose, found in dairy products.

Lactobacilli: Bacteria in the bowel and in probiotics. They are thought to be healthy bacteria.

Lactose: The sugar found in dairy products.

Laparoscopy: A surgical procedure in which a scope is placed through a small cut in the abdominal wall to visualize the abdominal cavity and pelvis.

Large intestine: See *colon*.

Laxative: A substance taken to cause a bowel movement.

LDL (low-density lipoprotein): A lipoprotein that transports cholesterol in the blood that is considered the bad cholesterol associated with increased heart disease.

Lower esophageal sphincter (LES): The high-pressure zone in the lower esophagus between the esophagus and stomach that prevents acid from backing up into the esophagus.

Lumen: The central portion (opening) of the tubelike structure of the digestive tract from mouth to anus.

Lupron (leuprolide): A synthetic version of the body's gonadotropin-releasing hormone that overstimulates the natural hormones in the body so that they shut down and decrease estrogen production and thereby decrease the stimulus for endometriosis.

Menstruation: The time during which a woman bleeds (her period) in her menstrual cycle.

Microbiome: The living organisms that are contained in the lumen of the intestines.

Motility: Movement of food through the digestive tract via muscle contractions of the walls of the digestive tract structures.

MRI scan (magnetic resonance imaging): A radiology test using a magnetic field to visualize parts of the body.

Multiple sclerosis: An autoimmune condition that affects the brain and spinal cord and causes problems with nerve function.

NSAIDs (nonsteroidal anti-inflammatory drugs): Medications that are not steroids used to decrease inflammation.

Oligofructose: A sugar considered a prebiotic.

Osteopenia: Mild loss of calcium in the bone.

Osteoporosis: Loss of calcium in the bone that makes it brittle and puts it at risk of breaking.

OTC (over the counter): A drug available without a prescription.

Pathogen: An organism (bacterium, virus, yeast) that causes disease.

PCP: Primary care physician.

pH: A measurement of acidity. A pH value of 7 is neutral, with lower numbers more acidic and higher numbers more basic.

PMS (premenstrual syndrome): A cluster of physical, psychological and emotional symptoms occurring shortly before the onset of menstruation.

PPI (proton pump inhibitor): A drug that blocks the proton pump in the stomach, preventing acid production.

Prebiotics: Nutrients (food) for healthy bacteria.

Probiotics: Living organisms thought to have good effects on one's health.

Progestin: A synthetic hormone with effects similar to progesterone.

Psoriasis: An autoimmune skin condition that causes thick red skin and often arthritis.

Rectocoele: An outpouching of the rectum that when anterior pushes forward into the vaginal wall.

Rectum: The last part of the colon, attached to the anus.

Rheumatoid arthritis: An autoimmune condition with inflammation of the joints.

Saccharomycetes boulardii: A type of yeast that is used as a probiotic.

Sigmoid colon: The twisty part of the colon in the lower abdomen, between the descending colon and rectum.

Sigmoidoscopy: An examination of the lining of the lower part of the colon—the rectum, the sigmoid colon and usually the descending colon—with a tube (that could be flexible or rigid) with a light.

Sitz-marker capsule: A capsule containing twenty-four radio-opaque markers that is used to evaluate the movement of the colon.

Sjögren's syndrome: An autoimmune disorder that causes dry eyes and mouth.

Small intestine: The part of the digestive tract in which most of the digestion of food and absorption of nutrients into the body takes place.

Spastic colon: An old term meaning a condition in which there is pain (spasms) in the colon.

SSRI (selective serotonin reuptake inhibitor): An antidepressant drug that acts by blocking reuptake of serotonin into the nerves so that more of this neurotransmitter is available to act on the brain.

Stomach: The digestive tract organ that makes acid and starts the process of digestion of the food.

Streptococcus: A type of bacterium.

Suppository: A medication-containing capsule that is inserted into the rectum or vagina.

Systemic lupus erythematosus: An autoimmune disease in which antibodies are made against the person's own DNA, or genetic material.

Tissue transglutaminase: A substance (enzyme) that joins cells together in the body. Antibodies against this enzyme are important in celiac disease.

Tricyclic antidepressants: A class of medications used to treat depression and helpful for the treatment of IBS.

Upper endoscopy: A scope with a camera that directly visualizes (and forceps inserted through it can sample) the esophagus, stomach and duodenum.

Virus: A very small infection-causing organism that requires the host's own living cells to enable it to multiply and produce more viruses.

Yeast: A single-celled organism that reproduces by budding or division.

Zenker's diverticulum: An outpouching of the upper portion of the esophagus.

GI TESTS

A good physical examination by your doctor is important. However, it may not be enough to diagnose your condition. Below are tests that are helpful. Descriptions of the tests are provided, and the symptoms or conditions for which they are used are discussed.

They are rated for:

Convenience and comfort:

☺ = easy to do, no preparation, minimal discomfort

☻ = some inconvenience with the test, and/or easy preparation, and/or mild discomfort, embarrassment or ingestion of bad-tasting material

☹ = may require sedation or anesthesia, and/or extensive preparation, and/or moderate discomfort

Expense:

Includes hospital/facility and physician fees. These vary from facility to facility and from physician to physician. When a sample of tissue is removed and analyzed under a microscope, additional costs are incurred. Costs are relative and are not adjusted for the amount allowed by the insurance company, which is usually less than that billed to an individual. The physician fee is a small portion of the total (and could be as little as one in ten dollars charged). The amount the insurance companies allow facilities to collect for the procedures differs not only by the plan but also by the negotiated rate by the institution or facility. As a consumer, it's important that you clarify what part of the expense you will be expected to pay, even if you have insurance coverage. Furthermore, if you are personally responsible for the charges, you should see if you can negotiate a price before your procedure. If you have no insurance for the procedure, see if the charges can be reduced to the lowest amount reimbursed by insurance, and if need be, you can arrange a payment plan ahead of time. If the facility where you are to have the procedure won't do anything for you, perhaps another one will. Don't be afraid to discuss it with your physician and the institution or hospital. Relative costs (which may vary slightly in different locations) are indicated by:

$

$$

$$$

$$$$

$$$$$

Likely need for insurance approval:

Y = yes. Make sure that you have it before the test, or you could be stuck with a hefty bill.

N = no. Insurance is not usually needed.

Abdominal X-ray—☺, $, N. A plain X-ray done with you lying down and/or standing up. The test is helpful to see if there is a bowel obstruction or to get a very rough idea if there is an abnormality in the bowel, such as a big loop. It's not an especially sensitive test.

Sitz-marker study—☺, $, N. A sitz-marker capsule contains twenty-four nonabsorbable markers that can be seen with an X-ray. Plain abdominal X-rays are done five and often seven days after ingestion of the capsule (occasionally two capsules are ingested a day apart). This gives a rough idea of whether there's a problem with motility in the colon. If you ingest one capsule, all but perhaps two markers should be evacuated in five days.

Defogram—☺, $, N. An X-ray study in which barium is put into your rectum and sigmoid colon through a tube inserted into your rectum, and barium paste is put into your vagina. You have to evacuate the barium while X-rays are taken. This test gives valuable information on why someone may be constipated. It can be very embarrassing for some women.

Abdominal ultrasound—☺, $$, N. A painless study using a small probe on the outside of your abdomen, in which sound waves are used to image abdominal structures. It is particularly good for looking at the gallbladder and the bile ducts.

Barium swallow (esophagus study)—☺/☹, $, N. An X-ray test in which barium (a chalky white substance, sometimes flavored) is swallowed to examine the esophagus for problems.

Video swallow—☺/☹, $$, N. Like the barium swallow test. However, this test is done in conjunction with speech therapists to evaluate swallowing in the throat and upper esophagus. Different consistencies of liquids and different solids are evaluated. Bread soaked in barium or barium tablets may be swallowed to evaluate swallowing problems. It may be done with the barium swallow.

Upper GI with or without small bowel follow-through—☺/☹, $$, N. An X-ray test in which barium is swallowed to examine the lower esophagus, stomach and duodenum. The small bowel follow-through evaluates the complete small intestine.

Gastric emptying study—☺, $$$, N. A nuclear medicine test in which you are given liquid and low-fat egg whites or another solid that is labeled with a marker, technetium-99m sulfur colloid, that is seen on an X-ray scan. It evaluates how fast the food moves out of your stomach. The residual amount remaining is considered normal if the residual is 37 to 90 percent at one hour, 30 to 60 percent at two hours and 0 to 10 percent at four hours. This test may be useful to assess why someone has persistent nausea, upper abdominal distension, pain after eating or persistent reflux symptoms in spite of adequate medication. It may also be useful to look for very fast emptying of the stomach.

Abdominal CAT scan with or without small bowel exam (enterography)—☺, $$$$, Y. A test, utilizing X-rays, to visualize the structures in your abdomen. You will be lying down under a scanner. To get an adequate view of your intestinal tract, you will drink dilute barium or water-soluble contrast. Usually an intravenous injection of an iodine-containing contrast to evaluate the blood vessels is also done. If you are allergic to iodine-containing substances, be sure to tell the radiologist, as you should not get the intravenous contrast. If you have severe kidney disease, the intravenous contrast cannot be used. This test is usually done with the pelvis CT if general abdominal pain is being evaluated. The enterography particularly focuses on the small bowel. This test sometimes gives more information than the small bowel follow-through, because it can examine structures and problems around and outside of the bowel. It is used for the evaluation of abdominal pain, weight loss, inflammatory bowel disease and possible disease of the other organs in the vicinity of the abdomen, such as the liver, pancreas, gallbladder or kidney.

Pelvis CAT scan with contrast—☺, $$$, Y. Identical to the abdominal CT, except that the pelvis and pelvic organs are examined.

Virtual colonoscopy—☺/☹, $$$$, Y. A CAT scan in which a special technique is used to evaluate the colon for polyps and cancers. It requires a colonoscopy preparation to have adequate visualization of the wall of the colon and not have stool masquerade as a polyp. A small amount of oral dilute barium is usually used, along with a substance to stop movement in the bowel. At this time, if the test is done in an experienced center, studies show that for polyps or colon cancer greater than two-fifths of an inch (ten millimeters) it may be as good as a colonoscopy. For polyps six to nine millimeters in size, virtual colonoscopy may detect only about three in five polyps. How good it is to detect even smaller polyps remains to be determined. If a polyp is found, a colonoscopy will subsequently have to be done to remove the polyp. At this time, insurance does not routinely cover this procedure for colorectal cancer screening.

Abdominal MRI scan with or without small bowel exam (enterography)—☺, $$$$$, Y. Magnetic resonance imaging of the abdomen involves a magnetic field, not X-rays. You are lying down and moved into a tube that is surrounded by a magnet. It is noisy and can feel claustrophobic to some people. There are open scanners, but they do not provide the same detail as the regular, more narrow scanners. If you are very obese, a scan on a regular scanner cannot be done. Sometimes oral contrast is given. An MRI cannot be done if you have certain metal material in your body. Make sure you remove all jewelry before having the test. The MR enterography is an excellent way to evaluate the small bowel for disease. Oral contrast containing dilute Guriam is given. When gadolinium is given intravenously, it can usually distinguish active inflammation from chronic scarring, such as would be seen in Crohn's disease or other conditions. Giving a substance to decrease movement in the small intestine makes for a better test result. This test should be done in a center that is very experienced in this type of exam.

Pelvis MRI scan—☺, $$$$$, Y. Magnetic resonance imaging of the pelvis, done like the abdominal MRI scan. It evaluates pelvis diseases. It has been shown to be particularly useful in the diagnosis of endometriosis and perianal disease (to examine fissures, fistulae and local abscesses).

MRCP—☺, $$$$–$$$$$, Y. Magnetic resonance cholangiopancreatography, a noninvasive technique utilizing an MRI scan to evaluate the bile ducts (the tubes that carry bile, made in the liver) inside and outside the liver and the pancreatic duct. It is especially good at evaluating the liver, gallbladder and pancreas. Small stones in the ducts may be missed, but this test is over 90 percent accurate for evaluating abnormal ducts. Sometimes oral contrast is given.

Breath tests—☺/☺, $, N. Tests that measure gas that you breathe out in a tube to evaluate poor absorption of sugars, which indicates the presence of *Helicobacter pylori* or bacterial overgrowth. Lactose or fructose is ingested to evaluate malabsorption of those compounds. If you don't absorb the sugars, they can cause intestinal gas, bloating and pain due to the release of hydrogen gas and the production of fatty acids by the bacteria. Lactulose is ingested to look for bacterial overgrowth in the small intestines (hydrogen and methane gases are measured). For *H. pylori* testing, a labeled carbon isotope, C-13 or C-14, is given by mouth (C-13 is not radioactive), and the bacteria incorporate this substance into carbon dioxide when they break down urea. The carbon dioxide is measured. The breath test for *H. pylori* can be used to look for active disease and assess that treatment of the bacteria was successful.

Upper endoscopy—☺, $$$–$$$$, Y. A scope with a camera directly visualizes, and forceps inserted through it can sample, the esophagus, stomach and duodenum. The back of your throat is anesthetized with a spray or gargle. Most people want to have conscious sedation or, more recently, propofol sedation. This test is useful to evaluate GERD that is not responding to therapy, trouble swallowing, possible ulcer disease, gastritis, celiac disease, and causes for bleeding, diarrhea or anemia. Often bleeding can be treated at the time of the procedure. If you have sedation, you should have nothing by mouth for four hours before the test, except for necessary medications (check with your GI doctor).

Enteroscopy—☺, $$$$, Y. A scope with a camera directly visualizes, and forceps inserted through it can sample, the small intestine. The test is done like the upper endoscopy, but the scope is longer and can be advanced farther into the small intestine to look for sites of bleeding and inflammation or possibly tumors. Depending on the type of test, a preparation may or may not be needed. If you have sedation, you should have nothing by mouth for four hours before the test, except for perhaps your necessary medications (check with your GI doctor).

Colonoscopy—☺/☹, $$$–$$$$$, Y/N. A scope with a camera directly visualizes the large intestine (colon) and possibly the last part of the small intestine (terminal ileum). Polyps can be removed, samples of tissue taken, and bleeding sources often stopped by instruments inserted through the tube. Sedation is desired by most people. The colon has to be cleaned out before the test in order to completely visualize the lining of the colon. This is done by a diet that is low in fiber for a few days before the test and then a clear liquid diet the day before the test. Laxatives are given the day before the test. In determining the exact type of laxative to prescribe, your doctor may take into consideration any chronic condition that you have or your ability to take the preparation. Possible colon clean-out techniques include drinking a gallon of a slightly salty solution (polyethylene glycol-containing liquid), magnesium citrate or sodium phosphosoda pills. Liquid sodium phosphate is no longer used for colonoscopy preps, as it can cause problems with the kidneys. At this time, the sodium phosphosoda pills have only rarely been shown to cause a problem with the kidneys. Make sure you drink a lot of fluid (electrolyte solution is good) to stay hydrated during your prep. If you have sedation, you should have nothing by mouth for four hours before the test, except for necessary medications (check with your GI doctor).

Flexible sigmoidoscopy—☺, $$, N. A scope with a camera directly visualizes the left side of the colon. Two Fleet enemas are used before the test to clean out the lower part of the colon. This is not a good test to evaluate bleeding or look for colon polyps or cancer, because as much as two-thirds of the colon is not visualized. It is often used to assess whether colitis may be active or to follow up on a previous finding. It's sometimes used for younger people experiencing minor bleeding. Sedation is usually not used.

Anoscopy—☺, $, N. A short metal or plastic tube with an attached light or an accompanying flashlight visualizes the anus. This can be done in the office with no preparation. It is often used to visualize hemorrhoids in order to see if they are bleeding or to visualize an anal fissure.

ERCP (endoscopic retrograde cholangiopancreatography)—☹, $$$$$, Y. A scope (tube) with a camera is inserted through the mouth and down into the duodenum, where a small tube is inserted through the scope into the bile ducts or pancreatic duct. Sedation is necessary. It is used for diagnosing and treating stones, tumors and the narrowing of the bile ducts and pancreatic duct. Except for perhaps your necessary medications (check with your GI doctor), you should have nothing by mouth for four hours before the test.

Esophageal motility test—☺, $$, N. A small tube inserted through the nose measures the pressures in the esophagus and assesses esophageal movement. You will be asked to drink water during the test and do multiple swallows. Most tubes have multiple holes so that readings of pressures from different areas can be taken at the same time.

Esophageal 24-hour pH probe with or without impedance study—☺/☹, $–$$$, N. A small tube inserted through the nose and advanced into the lower part of the esophagus measures acid and fluid that comes back into the esophagus (measured by a change in esophageal wall resistance during the impedance study). This test is used to evaluate reflux symptoms that persist after you have been on adequate antacid treatment, a chronic cough that could be due to reflux, atypical chest pain, and symptoms after surgery for reflux. The test may be done on or off therapy for acid, depending on your doctor's request. The tube can be uncomfortable for some people. You can go about your day otherwise normally. There is no need for you to stay home from work, though your day will likely be filled with questions from your colleagues! Usually you will be asked to have nothing by mouth for four to six hours before the test.

Wireless (Bravo) capsule to measure the pH—☺/☹, $$$, Y. A capsule attached to the esophageal mucosa at the time of upper endoscopy to measure the acid that refluxes into the esophagus at that one point. The advantages are that you don't have to have a tube down your nose to measure the acid, and if desired, recording can go on for two to four days. The disadvantages are that as many as one in four people have chest pain from the capsule. It eventually falls off the esophagus wall and is eliminated with the stool.

Enteric capsule study to look at the small intestine or elsewhere in the GI tract—☺/☹, $$$$$, Y. A swallowed capsule that takes pictures of the small intestine and transmits the pictures to a recording device. Its best utility is for evaluating intestinal bleeding and suspected or active Crohn's disease in the small intestine. It is being evaluated for its use to detect polyps in the colon. If you have a narrowing in the intestines, it should not be used, as it can possibly get stuck. Before swallowing the capsule, many GI doctors want you to take a colonoscopy prep in order to make sure that the lining of the small intestine is well seen.

Rectal motility—☺, $$–$$$, N. Measures the pressure in your anal sphincter at rest, and with squeezing and straining to evacuate, and evaluates when you feel a balloon distended in your rectum and if you are able to evacuate the balloon. A small probe and a balloon are inserted in your rectum. The balloon is distended until you feel it. This test evaluates rectal and anal sphincter function as a cause for constipation or fecal incontinence.

Rectal ultrasound—☺, $$$, Y/N. An ultrasound device on the end of a scope is inserted into the rectum. Ultrasound pictures of the surrounding area are made. It is helpful for evaluating the anal sphincter, looking for endometriosis and evaluating local tumors. An enema is required before the test.

Endoscopic ultrasound—☺, $$$$$, Y. An ultrasound device on the end of a scope is inserted through the mouth into the esophagus, stomach or perhaps small intestine. Ultrasound pictures of the surrounding area are made. This is helpful for evaluating tumors, local lymph nodes or the pancreas. No preparation is needed. If you have sedation, you should have nothing by mouth for six hours before the test, except for necessary medications (check with your GI doctor).

RESOURCES

GENERAL BOOKS

Yoshida, Cynthia M. *No More Digestive Problems*. With Deborah Kotz. New York: Bantam Dell, 2004.

Greenberger, Phyllis, *The Savvy Woman Patient: How and Why Sex Differences Affect Your Health*. With Jennifer Wider. Herndon, VA: Capital Books Inc, 2006.

Green, Peter H. R. and Rory Jones. *Celiac Disease: A Hidden Epidemic*. New York: HarperCollins, 2006.

Gottschall, Elaine. *Breaking the Vicious Cycle: Intestinal Health Through Diet*. Baltimore, ON: Kirkton Press, 1994.

Greenberger, Norton J. *4 Weeks to Healthy Digestion: A Harvard Doctor's Proven Plan for Reducing Symptoms of Diarrhea, Constipation, Heartburn, and More*. With Roanne Weisman. New York: McGraw-Hill, 2009.

WEBSITES

General Information

Information for patients from the American Gastroenterology Association
http://www.gastro.org/

Information for patients from the American Society for Gastrointestinal Endoscopy
http://www.asge.org/patientInfoIndex.aspx?id=1022

Information for patients from the American College of Gastroenterology
http://www.gi.org/patients/

Information on health issues of concern for women
http://www.womenshealth.gov/FAQ

Information on gastrointestinal diseases and ongoing clinical trials from
the National Digestive Diseases Information Clearinghouse (NDDIC)
http://digestive.niddk.nih.gov/

General source for health information, including digestive diseases,
from the Mayo Clinic
http://www.mayoclinic.com/

Information on complementary and alternative medicine from the
National Center for Complementary and Alternative Medicine (NCCAM)
http://nccam.nih.gov/

Free online newsletter on health issues from Harvard Medical School
http://www.health.harvard.edu/healthbeat/

Nutrition and Diet Information

http://www.tufts.edu/med/nutrition-infection/hiv/health_fiber.html

http://www.nhlbi.nih.gov/hbp/prevent/h_eating/h_eating.htm

http://www.hsph.harvard.edu/nutritionsource/index.html

http://www.mypyramid.gov/pyramid/index.html

http://www.DietaryGuidelines.gov

http://www.fsis.usda.gov/FactSheets/

http://www.healthierus.gov/dietaryguidelines

http://www.gicare.com/Diets/Diets-main.aspx#

Product Information

(This does not constitute an endorsement of a specific product.)

Flatulence odor control products:

http://www.flat-d.com/

http://www.under-tec.com/index.php

http://www.myshreddies.com/flatulence-incontinence-underwear/

Organizations with Helpful Patient Information (many with Internet links to other organizations of interest)

Celiac Disease:

Celiac Disease Foundation (CDF)
13251 Ventura Boulevard, Suite 1
Studio City, CA 91604
Phone: 818-990-2354
http://www.celiac.org

Celiac Sprue Association/USA, Inc. (CSA)
P.O. Box 31700
Omaha, NE 68131
Phone: 1-877-CSA-4CSA (1-877-272-4272)
http://www.csaceliacs.org

Gluten Intolerance Group of North America (GIG)
31214 124th Avenue SE
Auburn, WA 98092
Phone: 253-833-6655
http://www.gluten.net

National Foundation for Celiac Awareness (NFCA)
P.O. Box 544
Ambler, PA 19002-0544
Phone: 215-325-1306
http://www.CeliacCentral.org

Canadian Celiac Association
5170 Dixie Road, Suite 204
Mississauga, ON L4W 1E3
Canada
Phone: 905-507-6208
Toll-free: 1-800-363-7296
http://www.celiac.ca

Endometriosis:

Endometriosis Association
8585 N. 76th Place
Milwaukee, WI 53223
Phone: 414-355-2200
http://www.endometriosisassn.org

Irritable Bowel Syndrome and Intestinal Motility Problems:

International Foundation for Functional Gastrointestinal Disorders
P.O. Box 170864
Milwaukee, WI 53217-8076
Phone: 414-964-1799
Toll-free: 1-888-964-2001
http://www.iffgd.org

Association of Gastrointestinal Motility Disorders, Inc. (AGMD)
12 Roberts Drive
Bedford, MA 01730
Phone: 781-275-1300
http://www.agmd-gimotility.org

National Association for Continence (NAFC)
P.O. Box 1019
Charleston, SC 29402-1019
Phone: 843-377-0900
Toll-free: 1-800-BLADDER (1-800-252-3337)
http://www.nafc.org

Inflammatory Bowel Disease—Crohn's Disease and Colitis:

Crohn's & Colitis Foundation of America (CCFA)
386 Park Avenue South, 17th floor
New York, NY 10016
Phone: 212-685-3440
Toll-free: 1-800-932-2423
http://www.ccfa.org

Crohn's & Colitis Foundation of Canada
600-60 St. Clair Avenue East
Toronto, ON M4T 1N5
Canada
Phone: 416-920-5035
Toll-free: 1-800-387-1479
http://www.ccfc.ca

Stoma Care and Information:

United Ostomy Associations of America, Inc. (UOAA)
P.O. Box 66
Fairview, TN 37062-0066
Phone: 1-800-826-0826
http://www.uoaa.org

References

Introduction

Everhart, J. E., ed. "The Burden of Digestive Diseases in the United States." U.S. Department of Health and Human Services, Public Health Service, National Institutes of Health, National Institute of Diabetes and Digestive and Kidney Diseases. Washington, DC: U.S. Government Printing Office, 2008; NIH Publication No. 09-6443 [pp. 1–7].

Chapter 1: How Uncouth: Stomach Shame

Abraczinskas, D. and S.E. Goldfinger. "Intestinal Gas and Bloating." Topic updated August 25, 2009. http://www.uptodate.com.

Arhan, P., G. Devroede, B. Jehannin, M. Lanza, C. Faverdin, C. Dornic, B. Persoz, L. Tetreault, B. Perey, and D. Pellerin. "Segmental Colonic Transit Time." *Diseases of the Colon & Rectum* 24 (1981): 625–9.

Böll, Heinrich. *Group Portrait with Lady.* Translated by L. Vennewitz. New York: Avon Books, 1973.

Camilleri, M., D. Dubois, B. Coulie, M. Jones, P. J. Kahrilas, A. M. Rentz, A. Sonnenberg, et al. "Prevalence and Socioeconomic Impact of Upper Gastrointestinal Disorders in the United States: Results of the US Upper Gastrointestinal Study." *Clinical Gastroenterology and Hepatology* 3 (2005): 543–52.

Cani, P. D., N. M. Delzenne, J. Amar, R. Burcelin. "Role of Gut Microflora in the Development of Obesity and Insulin Resistance Following High-fat Diet Feeding." *Pathologie Biologie* 56 (2008): 305–9.

Cecil, J. E., R. Tavendale, P. Watt, et al. "An Obesity-associated FTO Gene Variant and Increased Energy Intake in Children." *New England Journal of Medicine* 359 (2008): 2558–66.

Collado, M. C., E. Isolauri, K. Laitinen, and S. Salminen. "Distinct Composition of Gut Microbiota during Pregnancy in Overweight and Normal-weight Women." *American Journal of Clinical Nutrition* 88 (2008): 894–9.

Dent, J., H. B. El-Serag, M. A. Wallander, and S. Johansson. "Epidemiology of Gastro-oesophageal Reflux Disease: A Systematic Review." *Gut* 54 (2005): 710–7.

Gerson, L. B. "Effect of Heartburn on Life Expectancy." *Gastroenterology* 134 (2008): 2182–4.

Hutson, W. R., R. L. Roehrkasse, and A. Wald. "Influence of Gender and Menopause on Gastric Emptying and Motility." *Gastroenterology* 96 (1989): 11–7.

Kalliomaki, M., M. C. Collado, S. Salminen, et al. "Early Differences in Fecal Microbiota Composition in Children May Predict Overweight." *American Journal of Clinical Nutrition* 87 (2008): 534–38.

Karlstadt, R. G., D. L. Hogan, and A. Foxx-Orenstein. "Normal Physiology of the Gastrointestinal Tract and Gender Differences." In *Principles of Gender-Specific Medicine,* edited by M. J. Legato, 377–95. San Diego: Elsevier Academic Press, 2004.

Legato, M. J., ed. *Principles of Gender-Specific Medicine.* San Diego: Elsevier Academic Press, 2004.

Ley, R. E., F. Bäckhed, P. Turnbaugh, C. A. Lozupone, R. D. Knight, and J. I. Gordon. "Obesity Alters Gut Microbial Ecology." *The Proceedings of the National Academy of Sciences* 102 (2005): 11070–5.

McCrea, G. L., C. Miaskowski, N. A. Stotts, L. Macera, S. M. Paul, and M. G. Varma. "Gender Differences in Self-reported Constipation Characteristics,

Symptoms, and Bowel and Dietary Habits among Patients Attending a Specialty Clinic for Constipation." *Gender Medicine* 6 (2009): 259–70.

Persels, J., and R. Ganim, eds. *Fecal Matters in Early Modern Literature and Art: Studies in Scatology.* Aldershot, Hants, England: Ashgate, 2004.

Rex, D. K., D. A. Johnson, J. C. Anderson, P. S. Schoenfeld, C. A. Burke, and J. M. Inadomi. "American College of Gastroenterology Guidelines for Colorectal Cancer Screening 2008." *American Journal of Gastroenterology* 104 (2009): 739–50.

Sabbath, D., and M. Hall. *End Product: The First Taboo.* New York: Urizen Books, 1977.

Sanjeevi, A., and D. F. Kirby. "The Role of Food and Dietary Intervention in the Irritable Bowel Syndrome." *Practical Gastroenterology* 32 (2008): 33–42.

Shaheen, N., and D. F. Ransohoff. "Gastroesophageal Reflux, Barrett Esophagus, and Esophageal Cancer: Scientific Review." *Journal of the American Medical Association* 287 (2002): 1972–81.

Tilg, H., A. R. Moschen and A. Kaser. "Obesity and the Microbiota." *Gastroenterology* 36 (2009): 1476–83.

Turnbaugh, P. J., M. Hamady, T. Yatsunenko, B. L. Cantarel, A. Duncan, R. E. Ley, M. L. Sogin, et al. "A Core Gut Microbiome in Obese and Lean Twins." *Nature* 457 (2009): 480–4.

Turnbaugh, P. J., V. K. Ridaura, J. J. Faith, F. E. Rey, R. Knight, and J. I. Gordon. "The Effect of Diet on the Human Gut Microbiome: A Metagenomic Analysis in Humanized Gnotobiotic Mice." *Science Translational Medicine* 1 (2009): 1–10.

Harvard Medical School. "What is a Healthy Bowel Movement? The Characteristics of Feces Can Offer Clues to Health Problems, Digestive and Otherwise." Harvard Health Letter, July 2004. www.health.harvard.edu.

Zhang, H., J. K. DiBaise, A. Zuccolo, D. Kudrna, M. Braidotti, Y. Yu, P. Parameswaran, et al. "Human Gut Microbiota in Obesity and after Gastric Bypass." *PNAS The Proceedings of the National Academy of Sciences* 106 (2009): 2365–70.

http://nccam.nih.gov/health/probiotics/.

Chapter 2: Endometriosis and Feminine GI Troubles: Symptoms Every Woman Should Understand

Bailey, M. T., and C. L. Coe. "Endometriosis is Associated with an Altered Profile of Intestinal Microflora in Female Rhesus Monkeys." *Human Reproduction* 17 (2002): 1704–8.

Barnes, W., S. Waggoner, G. Delgado, K. Maher, R. Potkul, J. Barter, and S. Benjamin. "Manometric Characterization of Rectal Dysfunction Following Radical Hysterectomy." *Gynecologic Oncology* 42 (1991): 116–9.

Bazot, M., E. Darai, R. Hourani, I. Thomassin, A. Cortez, S. Uzan, and J. N. Buy. "Deep Pelvic Endometriosis: MR Imaging for Diagnosis and Prediction of Extension of Disease." *Radiology* 232 (2004): 379–89.

Bertone-Johnson, E. R., S. E. Hankinson, A. Bendich, S. R. Johnson, W. C. Willett, and J. E. Manson. "Calcium and Vitamin D Intake and Risk of Incident Premenstrual Syndrome." *Archives of Internal Medicine* 165 (2005): 1246–52.

Crosignani, P., D. Olive, A. Bergqvist, and A. Luciano. "Advances in the Management of Endometriosis: an Update for Clinicians." *Human Reproductive Update* 12 (2006): 179–89.

Dias Jr. J. A., R. M. de Oliveira, and M. S. Abrao. "Antinuclear Antibodies and Endometriosis." *International Journal of Gynecology & Obstetrics* 93 (2006): 262–3.

Farquhar, C. "Endometriosis." *British Medical Journal* 334 (2007): 249–53.

Fjerbaek, A., and U. B. Knudsen. "Endometriosis, Dysmenorrhea and Diet—What is the Evidence? [Review]." *European Journal of Obstetrics & Gynecology and Reproductive Biology* 132 (2007): 140–7.

Forsgren, C., J. Zetterstrom, A. Lopez, J. Nordenstam, B. Anzen, and D. Altman. "Effects of Hysterectomy on Bowel Function: a Three-year, Prospective Cohort Study." *Diseases of the Colon & Rectum* 50 (2007); 1139–45.

Greene, R., P. Stratton, S. D. Cleary, M. L. Ballweg, and N. Sinaii. "Diagnostic Experience among 4,334 Women Reporting Surgically Diagnosed Endometriosis." *Fertility and Sterility* 91 (2009) 32–9.

Kennedy, S., A. Bergqvist, C. Chapron, T. D'Hooghe, G. Dunselman, R. Greb, L. Hummelshoj, A. Prentice, and E. Saridogan. "ESHRE Guideline for the Diagnosis and Treatment of Endometriosis." *Human Reproduction* 20 (2005): 2698–704.

Melin, A., P. Sparen, and A. Bergqvist. "The Risk of Cancer and the Role of Parity among Women with Endometriosis." *Human Reproduction* 22 (2007): 3021–6.

Roovers, J. P., J. G. van der Bom, and C. H. van der Vaart. "Hysterectomy Does Not Cause Constipation." *Diseases of the Colon & Rectum* 51 (2008): 1068–72.

Seaman, H. E., K. D. Ballard, J. T. Wright, and C. S. de Vries. "Endometriosis and its Coexistence with Irritable Bowel Syndrome and Pelvic Inflammatory Disease: Findings from a National Case-Control Study—Part 2." *BJOG: An International Journal of Obstretics and Gynaecology* 115 (2008): 1392–426.

Sinaii, N., S. D. Cleary, M. L. Ballweg, L. K. Nieman, and P. Stratton. "High Rates of Autoimmune and Endocrine Disorders, Fibromyalgia, Chronic Fatigue Syndrome and Atopic Diseases among Women with Endometriosis: A Survey Analysis." *Human Reproduction* 17 (2002): 2715–424.

Sinaii, N., S. D. Cleary, N. Younes, M. L. Ballweg, and P. Stratton. "Treatment Utilitzation for Endometriosis Symptoms: A Cross-sectional Survey Study of Lifetime Experience." *Fertility and Sterility* 87 (2007): 1277–86.

Sperber, A. D., C. B. Morris, L. Greemberg, S. I. Bangdiwala, D. Goldstein, E. Sheiner, Y. Rusabrov, et al. "Constipation Does Not Develop Following Elective Hysterectomy: A Prospective, Controlled Study." *Neurogastroenterology & Motility* 21 (2009): 18–22.

Sperber, A. D., C. B. Morris, L. Greemberg, S. I. Bangdiwala, D. Goldstein, E. Sheiner, Y. Rusabrov, et al. "Development of Abdominal Pain and IBS Following Gynecological Surgery: A Prospective, Controlled Study." *Gastroenterology* 134 (2008): 74–84.

Stanhope, C. R. "Chronic Constipation Following Simple Hysterectomy Is Rare [editorial; comment]." *Gynecologic Oncology* 42 (1991): 114–5.

Tsuchiya, M., T. Miura, T. Hanaoka, M. Iwasaki, H. Sasaki, T. Tanka, H. Nakao, et al. "Effect of Soy Isoflavones on Endometriosis: Interaction with Estrogen Receptor 2 Gene Polymorphism." *Epidemiology* 18 (2007): 402–8.

Vercellini, P., E. Somigliana, R. Daguati, P. Vigano, F. Meroni, and P. G. Crosignani. "Postoperative Oral Contraceptive Exposure and Risk of Endometrioma Recurrence." *American Journal of Obstetrics & Gynecology* 198 (2008): 504e1–5.

Whitehead, W. E., L. J. Cheskin, B. R. Heller, J. C. Robinson, M. D. Crowell, C. Benjamin, and M. M. Schuster. "Evidence for Exacerbation of Irritable Bowel Syndrome during Menses." *Gastroenterology* 98 (1990): 1485–9.

Wieser, F., M. Cohen, A. Gaeddert, J. Yu, C. Burks-Wicks, S. L. Berga, and R. N. Taylor. "Evolution of Medical Treatment for Endometriosis: Back to the Roots?" *Human Reproductive Update* 13 (2007): 487–99.

Wolf, J. L. "Bowel function." In *The Primary Care of Women*. 2nd ed. edited by K.J. Carlson and S.A. Eisenstat. St. Louis: Mosby-Yearbook, Inc., 2002. 133–141.

Lifetime Probability of Developing Cancer 2009. At www.cancer.org

What is CAM? [NCCAM Cam Basics] nccam.nih.gov/health/whatiscam/overview.htm

www.mayoclinic.com/.../ovarian-cancer/.../DSECTION=symptoms

http://www.womenshealth.gov/FAQ/premenstrual-syndrome.cfm#a

Chapter 3: "Do These Pants Come With an Elastic Waist?" The Truth about Gas, Bloating and Irritable Bowel Syndrome

Accarino, A., F. Perez, F. Azpiroz, S. Quiroga, and J. R. Malagelada. "Abdominal Distention Results from Caudo-ventral Redistribution of Contents." *Gastroenterology* 136 (2009): 1544–51.

Agrawal, A., L. A. Houghton, J. Morris, B. Reilly, D. Guynonets, N. G. Feuillerats, A. Schlumbergers, S. Jakobs, and P. J. Whorwell. "Clinical Trial: The Effects of a Fermented Milk Product Containing *Bifidobacterium lactis* DN-173 010 on Abdominal Distension and Gastrointestinal Transit in Irritable Bowel Syndrome with Constipation." *Alimentary Pharmacology & Therapeutics* 29 (2008): 104–14.

Alonso, C., M. Guilarte, M. Vicario, L. Ramos, Z. Ramadan, M. Antolin, C. Martinez, et al. "Maladaptive Intestinal Epithelial Responses to Life

Stress May Predispose Healthy Women to Gut Mucosal Inflammation." *Gastroenterology* 135 (2008): 163–72.

American College of Gastroenterology Task Force on Irritable Bowel Syndrome. "An Evidence-Based Systematic Review on the Management of Irritable Bowel Syndrome." *American Journal of Gastroenterology* 104, supplement 1 (2009): S1–S35.

Atkinson, W., S. Lockhart, P. J. Whorwell, B. Keevil, L. A. Houghton. "Altered 5-hydroxytryptamine Signaling in Patients with Constipation and Diarrhea-predominant Irritable Bowel Syndrome. *Gastroenterology* 130 (2006): 34–43.

Barrett, J. S., and P. R. Gibson. "Clinical Ramifications of Malabsorption of Fructose and other Short-chain Carbohydrates." *Practical Gastroenterology* 31 (2007): 51–65.

Bensoussan, A., N. J. Talley, M. Hing, R. Menzies, A. Guo, and M. Ngu. "Treatment of Irritable Bowel Syndrome with Chinese Herbal Medicine: A Randomized Controlled Trial." *Journal of the American Medical Association* 280 (1998): 1585–9.

Bijkerk, C. J., N. J. de Wit, J. W. M. Muris, P. J. Whorwell, J. A. Knotterus, and A. W. Hoes. "Soluble or Insoluble Fibre in Irritable Bowel Syndrome in Primary Care? Randomised Placebo Controlled Trial." *British Medical Journal* 339 (2009): b3154.

Cain, K. C., P. Headstrom, M. E. Jarrett, S. A. Motzer, H. Park, R. L. Burr, C. M. Surawicz, and M. M. Heitkemper. "Abdominal Pain Impacts Quality of Life in Women with Irritable Bowel Syndrome." *American Journal of Gastroenterology* 101 (2006): 124–32.

Camilleri, M., and E. A. Mayer. "Developing Irritable Bowel Syndrome Guidelines through Meta-analyses: Does the Emperor Really Have New Clothes?" *Gastroenterology* 137 (2009): 766–9.

Camilleri, M., S. McKinzie, I. Busciglio, P. A. Low, S. Sweetser, D. Burton, K. Baxter, M. Ryks, and A. R. Zinsmeister. "Prospective Study of Motor Sensory Psychologic, and Autonomic Functions in Patients with Irritable Bowel Syndrome." *Clinical Gastroenterology and Hepatology* 6 (2008): 772–81.

Chitkara, D. K., M. A. L. van Tilburg, N. Blois-Martin, and W. E. Whitehead. "Early Life Risk Factors that Contribute to Irritable Bowel Syndrome in Adults:

A Systematic Review." *American Journal of Gastroenterology* 103 (2008): 765–74.

Choung, R. S., G. R. Locke, A. R. Zinsmeister, C. D. Schleck, and N. J. Talley. "Epidemiology of Slow and Fast Colonic Transit Using a Scale of Stool Form in a Community." *Alimentary Pharmacology & Therapeutics* 26 (2007): 1043–50.

Drisko, J., B. Bischoff, M. Hall, and R. McCallum. "Treating Irritable Bowel Syndrome with a Food Elimination Diet Followed by Food Challenge and Probiotics." *Journal of the American College of Nutrition* 25 (2006): 514–22.

Drossman, A., M. Camilleri, E. A. Mayer, and W. E. Whitehead. "A Technical Review on Irritable Bowel Syndrome." *Gastroenterology* 123 (2002): 2108–31.

DuPont, H. L. "Postinfectious Irritable Bowel Syndrome: Clinical Aspects, Pathophysiology, and Treatment." *Practical Gastroenterology* 31 (September supplement, 2007) 18–24.

El-Serag, H. B., P. Pilgrim, and P. Schoenfeld. "Systematic Review: Natural History of Irritable Bowel Syndrome." *Alimentary Pharmacology & Therapeutics* 12 (2004): 861–70.

Eutamene, H., and L. Bueno. "Role of Probiotics in Correcting Abnormalities of Colonic Flora Induced by Stress." *Gut* 56 (2007) 1495–7.

Ford, A. C., D. Forman, A. G. Bailey, A. T. Axon, and P. Moayyedi. "Irritable Bowel Syndrome: a 10-Yr Natural History of Symptoms and Factors that Influence Consultation Behavior." *American Journal of Gastroenterology* 103 (2008): 1229–40.

Ford, A. C., N. J. Talley, P. S. Schoenfeld, E. M. Quigley, and P. Moayyedi. "Efficacy of Antidepressants and Psychological Therapies in Irritable Bowel Syndrome: Systematic Review and Meta-analysis." *Gut* 58 (2009): 367–78.

Ford, A. C., N. J. Talley, B. M. R. Spiegel, A. E. Foxx-Orenstein, L. Schiller, E. M. M. Quigley, et al. "Effect of Fibre, Antispasmodics and Peppermint Oil in the Treatment of Irritable Bowel Syndrome: Systematic Review of the Literature and Meta-analysis." *British Medical Journal* 337 (2008): a2313.

Ganiats, T. G., W. A. Norcross, A. L. Halverson, P. A. Burford, and L. A. Palinkas. "Does Beano Prevent Gas? A Double-blind Crossover Study of Oral Alpha-galactosidase to Treat Dietary Oligosaccharide Intolerance." *Journal of Family Practice* 39 (1994): 441–5.

Heitkemper, M. M., K. C. Cain, M. E. Jarrett, R. L. Burr, V. Hertig, and E. F. Bond. "Symptoms across the Menstrual Cycle in Women with Irritable Bowel Syndrome." *American Journal of Gastroenterology* 98 (2003): 420–30.

Heitkemper, M. M., and L. Chang. "Do Fluctuations in Ovarian Hormones Affect Gastrointestinal Symptoms in Women with Irritable Bowel Syndrome?" *Gender Medicine* 6 (2009): 152–67.

Heizer, W. D., S. Southern, and S. McGovern. "The Role of Diet in Symptoms of Irritable Bowel Syndrome in Adults: A Narrative Review." *Journal of the American Dietetic Association* 109 (2009): 1204–14.

Hussain, Z., and E. M. M. Quigley. "Systematic Review: Complementary and Alternative Medicine in the Irritable Bowel Syndrome." *Alimentary Pharmacology & Therapeutics* 23 (2006): 465–71.

Jiang, X., G. R. Locke 3rd, R. S. Choung, A. R. Zinsmeister, C. D. Schleck, and N. J. Talley. "Prevalence and Risk Factors for Abdominal Bloating and Visible Distention: A Population-based Study." *Gut* 57 (2008): 756–63.

Kane, S. V., K. Sable, and S. B. Hanauer. "The Menstrual Cycle and its Effects on Inflammatory Bowel Disease and Irritable Bowel Syndrome: A Prevalence Study." *American Journal of Gastroenterology* 93 (1998): 1867–72.

Kassinen, A., L. Krogius-Kurikka, H. Mäkivuokko, T. Rinttilä, I. Paulin, J. Corander, E. Malinen, J. Apajalahti, and A. Palva. "The Fecal Microbiota of Irritable Bowel Syndrome Patients Differs Significantly from that of Healthy Subjects." *Gastroenterology* 133 (2007): 24–33.

Langmead, L. and D. S. Rampton. "Review Article: Herbal Treatment in Gastrointestinal and Liver Disease—Benefits and Dangers." *Alimentary Pharmacology & Therapeutics* 14 (2001) 1239–52.

Lefy, R. L., K. R. Jones, W. E. Whitehead, S. I. Feld, N. J. Talley, and L. A. Corey. "Irritable Bowel Syndrome in Twins: Heredity and Social Learning Both Contribute to Etiology." *Gastroenterology* 121 (2001): 799–804.

Lembo, A. J. "The Clinical and Economic Burden of Irritable Bowel Syndrome." *Practical Gastroenterology* 31 (September supplement, 2007): 3–9.

Lembo, A. J. "Current Diagnostic Strategies and Pharmacologic Treatment Options for Irritable Bowel Syndrome." *Practical Gastroenterology* 31 (September supplement, 2007): 10–7.

Lembo, A. J., L. Conboy, J. M. Kelley, R. S. Schnyer, C. A. McManus, M. T. Quilty, C. E. Kerr, et al. "A Treatment Trial of Acupuncture in IBS Patients." *American Journal of Gastroenterology* 104 (2009): 1489–97.

Lembo, A. J., B. Neri, J. Tolley, D. Barken, S. Carroll, and H. Pan. "Use of Serum Biomarkers in a Diagnostic Test for Irritable Bowel Syndrome." *Alimentary Pharmacology & Therapeutics* 29 (2009): 834–42.

Levy, R. L., W. E. Whitehead, L. S. Walker, M. Von Korff, A. D. Feld, M. Garner, and D. Christie. "Increased Somatic Complaints and Health-care Utilization in Children: Effects of Parent IBS Status and Parent Response to Gastrointestinal Symptoms." *American Journal of Gastroenterology* 99 (2004): 2442–51.

Liu, J., M. Yang, Y. Liu, M. Wei, and S. Grimsgaard. "Herbal Medicines for Treatment of Irritable Bowel Syndrome [Review]." The Cochrane Library 2009. Issue 2. http://www.thecochranelibrary.com.

Longstreth, G. F., W. G. Thompson, W. D. Chey, L. A. Houghton, F. Mearin, and R. C. Spiller. "Functional Bowel Disorders." *Gastroenterology* 130 (2006): 1480–91.

Madisch, A., G. Holtmann, K. Plein, and J. Hotz. "Treatment of Irritable Bowel Syndrome with Herbal Preparations: Results of a Double-blind, Randomized, Placebo-controlled, Multi-centre Trial." *Alimentary Pharmacology & Therapeutics* 19 (2004): 271–9.

McFarland, L. V., and S. Dublin. "Meta-analysis of Probiotics for the Treatment of Irritable Bowel Syndrome." *World Journal of Gastroenterology* 14 (2008): 2650–61.

Moore, J. G., L. D. Jessop, and D. N. Osborne. "Gas-chromatographic and Mass-spectrometric Analysis of the Odor of Human Feces." *Gastroenterology* 93 (1987): 1321–9.

Naliboff, B. D., S. Berman, L. Chang, S. W. Derbyshire, B. Suyenobu, B. A. Vogt, M. Mandelkern, and E. A. Mayer. "Sex-related Differences in IBS Patients: Central Processing of Visceral Stimuli." *Gastroenterology* 124 (2003): 1738–47.

Nikfar, S., R. Rahimi, F. Rahimi, S. Derakhshani, and M. Abdollahi. "Efficacy of Probiotics in Irritable Bowel Syndrome: A Meta-analysis of Randomized, Controlled Trials." *Diseases of the Colon & Rectum* 51 (2008): 1775–80.

Ohge, H., J. K. Furne, J. Springfield, S. Ringwala, and M. D. Levitt. "Effectiveness of Devices Purported to Reduce Flatus Odor." *American Journal of Gastroenterology* 100 (2005) 397–400.

Park, M. I., and M. Camilleri. "Is There a Role of Food Allergy in Irritable Bowel Syndrome and Functional Dyspepsia? A Systemic Review." *Neurogastroenterology & Motility* 18 (2006): 595–607.

Parkes, G. C., J. Brostoff, K. Whelan, and J. D. Sanderson. "Gastrointestinal Microbiota in Irritable Bowel Syndrome: Their Role in its Pathogenesis and Treatment." *American Journal of Gastroenterology* 103 (2008): 1557–67.

Pimentel, M. "Bacteria and the Role of Antibiotics in Irritable Bowel Syndrome." *Practical Gastroenterology* 31 (September supplement, 2007): 25–32.

Rangnekar, A. S., and W. D. Chey. "The FODMAP Diet for Irritable Bowel Syndrome: Food Fad or Roadmap to a New Treatment Paradigm? (Selected Summaries)" *Gastroenterology* 137 (2009): 383–6.

Schneider, A., P. Enck, K. Streitberger, C. Weiland, S. Bagheri, S. Witte, H. C. Friedrich, W. Herzog, and S. Zipfel. "Acupuncture Treatment in Irritable Bowel Syndrome." *Gut* 55 (2006) 649–54.

Shanahan, F. "Irritable Bowel Syndrome: Shifting the Focus toward the Gut Microbiota [editorial]." *Gastroenterology* 133 (2007): 340–2.

Shepherd, S. J., F. C. Parker, J. G. Muir, and P. R. Gibson. "Dietary Triggers of Abdominal Symptoms in Patients with Irritable Bowel Syndrome: Randomized Placebo-controlled Evidence." *Clinical Gastroenterology and Hepatology* 6 (2008): 765–71.

Simren, M. "Bloating and Abdominal Distention: Not so Poorly Understood Anymore! [editorial]." *Gastroenterology* 136 (2009): 1487–505.

Spiller, R., and K. Garsed. "Postinfectious Irritable Bowel Syndrome." *Gastroenterology* 136 (2009): 1979–88.

Suarez, F. L., J. Springfield, and M. D. Levitt. "Identification of Gases Responsible for the Odour of Human Flatus and Evaluation of a Device Purported to Reduce This Odour." *Gut* 43 (1998): 100–4.

Talley, N. J. "Genes and Environment in Irritable Bowel Syndrome: One Step Forward." *Gut* 55 (2006): 1694–96.

Tillisch, K. "Complementary and Alternative Medicine for Functional Gastrointestinal Disorders." *Clinical Medicine* 7 (2007): 224–7.

Tillisch, K. "Complementary and Alternative Medicine for Functional Gastrointestinal Disorders." *Gut* 55 (2006): 593–96.

Videlock, E. J., M. Adeyemo, A. Licudine, M. Hirano, G. Ohning, M. Mayer, E. A. Mayer, and L. Chang. "Childhood Trauma is Associated with Hypothalamic-pituitary-adrenal Axis Responsiveness in Irritable Bowel Syndrome." *Gastroenterology* 137 (2009): 1954–62.

Whorwell, P. J., L. Altringer, J. Morel, Y. Bond, D. Charbonneau, L. O. Mahony, B. Kiely, F. Shanahan, and E. M. M. Quigley. "Efficacy of an Encapsulated Probiotic *Bifidobacterium infantis* 35624 in Women with Irritable Bowel Syndrome." *American Journal of Gastroenterology* 101 (2006): 1581–90.

Williams, E. A., J. Stimpson, D. Wang, S. Plummer, I. Garaiova, M. E. Barker, and B. M. Corfe. "Clinical Trial: A Multistrain Probiotic Preparation Significantly Reduces Symptoms of Irritable Bowel Syndrome in a Double-blind Placebo-controlled Study." *Alimentary Pharmacology & Therapeutics* 29 (2008) 97–103.

http://www.healthhype.com/fodmap-diet-foods-to-avoid-in-ibs-bowel-disorders-with-bloating-and-gas.html

Chapter 4: When You Know Every Bathroom in Town: Diarrhea

"AGA Institute Medical Position Statement on the Diagnosis and Management of Celiac Disease." *Gastroenterology* 131 (2006): 1977–80.

"American Gastroenterological Association (AGA) Institute Technical Review on the Diagnosis and Management of Celiac Disease." *Gastroenterology* 131 (2006): 1981–2002.

Austin, G. L., C. B. Dalton, Y. Hu, C. B. Morris, J. Hankins, S. R. Weinland, E. C. Westman, W. S. Yancy Jr., and D. A. Drossman. "A Very Low-carbohydrate Diet

Improves Symptoms and Quality of Life in Diarrhea-predominant Irritable Bowel Syndrome." *Clinical Gastroenterology and Hepatology* 7 (2009): 706–8.

Barrett, J. S., and P. R. Gibson. "Clinical Ramifications of Malabsorption of Fructose and Other Short-chain Carbohydrates." *Practical Gastroenterology* 31 (2007): 51–65.

Bleday, R., and E. Breen. "Treatment of Hemorrhoids." www.update.com. Topic updated April 14, 2009.

Dimeo, F. C., J. Peters, and H. Guderian. "Abdominal Pain in Long Distance Runners: Case Report and Analysis of the Literature." *British Journal of Sports Medicine* 38 (2004): e24.

Diop, L., S. Guillou, and H. Durand. "Probiotic Food Supplement Reduces Stress-induced Gastrointestinal Symptoms in Volunteers: A Double-blind, Placebo-controlled, Randomized Trial." *Nutrition Research* 28 (2008): 1–5.

DuPont, H. L. "Clinical Practice. Bacterial Diarrhea." *New England Journal of Medicine* 361 (2009): 1560–69.

Dupont, H. L. "Travellers' Diarrhoea: Contemporary Approaches to Therapy and Prevention." *Drugs* 66 (2006): 303–14.

Engelbrektson, A., J. R. Korzenik, A. Pittler, M. E. Sanders, T. R. Klaenhammer, G. Leyer, and C. L. Kitts. "Probiotics to Minimize the Disruption of Faecal Micobiota in Healthy Subjects Undergoing Antibiotic Therapy." *Journal of Medical Microbiology* 58 (2009): 663–70.

Everhart, J. E., ed. "The Burden of Digestive Diseases in the United States." U.S. Department of Health and Human Services, Public Health Service, National Institutes of Health, National Institute of Diabetes and Digestive and Kidney Diseases. Washington, DC: U.S. Government Printing Office, 2008; NIH Publication No. 09-6443 [pp. 9–10].

ESPGHAN Committee on Nutrition. "Complementary Feeding: A Commentary by the ESPGHAN Committee on Nutrition." *Journal of Pediatric Gastroenterology and Nutrition* 46 (2008): 99–110.

Fasano, A., I. Berti, T. Gerarduzzi, N. Tarcisio, R. B. Colletti, S. Drago, Y. Elitsur, et al. "Prevalence of Celiac Disease in At-risk and Not-at-risk Groups in the United States: A Large Multicenter Study." *Archives of Internal Medicine* 163 (2003): 286–92.

Glickman, M. S., and M. B. Goldberg. "Microbiology and Epidemiology of Shigella Infection." www.uptodate.com. Topic updated January 15, 2007.

Green, P. H. "Mortality in Celiac Disease, Intestinal Inflammation, and Gluten Sensitivity." *Journal of the American Medical Association* 302 (2009): 1225–6.

Green, Peter R., and R. Jones. *Celiac Disease: A Hidden Epidemic.* New York: HarperCollins, 2006.

Hlywiak, K. H. "Hidden Sources of Gluten." *Practical Gastroenterology* 32 (2008): 27–39.

Imhoff, B., D. Morse, B. Shiferaw, M. Hawkins, D. Vugia, S. Lance-Parker, J. Hadler, et al. "Burden of Self-reported Acute Diarrheal Illness in FoodNet Surveillance Areas, 1998–1999." *Clinical Infectious Diseases* 38 (April supplement 3, 2008): S219–26.

Jatla, M., P. A. Bierly, K. H. Hlywiak, J. Autodore, and R. Verma. "Overview of Celiac Disease: Differences Between Children and Adults." *Practical Gastroenterology* 32 (2008): 18–34.

Karasick, S., D. Karasick, and S. R. Karasick. "Functional Disorders of the Anus and Rectum: Findings on Defecography." *American Journal of Roentgenology* 160 (1993): 777–82.

Kaukinen, K., J. Partanen, M. Mäki, and P. Collin. "HLA-DQ Typing in the Diagnosis of Celiac Disease." *American Journal of Gastroenterology* 97 (2002): 695–9.

Kelly, C. "Diagnosis of Celiac Disease." www.uptodate.com. Topic updated September 18, 2009.

Lanzini, A., F. Lanzarotto, V. Villanacci, A. Mora, S. Bertolazzi, D. Turini, G. Carella, et al. "Complete Recovery of Intestinal Mucosa Occurs Very Rarely in Adult Celiac Patients Despite Adherence to Gluten-free Diet." *Alimentary Pharmacology Therapeutics* 29 (2009): 1299–1308.

Leder, K., and P. F. Weller. "Cryptosporidiosis." www.uptodate.com. Topic updated May 22, 2009.

Leffler, D. A., and C. P. Kelly. "Celiac Disease: What the Last Few Years Have Taught Us." In *Advances in Digestive Disease,* edited by Colin W. Howden, 49–58. Bethesda, MD: AGA Institute Press, 2007.

Ludvigsson, J. F., S. M. Montgomery, A. Ekbom, L. Brandt, and F. Granath. "Small-Intestinal Histopathology and Mortality Risk in Celiac Disease." *Journal of the American Medical Association* 302 (2009): 1171–218.

Majowicz, S. E., K. Dore, J. A. Flint, V. L. Edge, S. Read, M. C. Buffett, S. McEwen, et al. "Magnitude and Distribution of Acute, Self-reported Gastrointestinal Illness in a Canadian Community." *Epidemiology Infection* 132 (2004): 607–17.

McFarland, L. V. "Meta-analysis of Probiotics for the Prevention of Traveler's Diarrhea." *Travel Medicine and Infectious Disease* 5(2007): 97.

Montgomery, R. K., R. J. Grand, and H. A. Büller. "Lactose Intolerance." www.uptodate.com. Topic updated August 26, 2008.

Morton, D. P., and R. Callister. "Characteristics and Etiology of Exercise-related Transient Abdominal Pain." *Medicine & Science in Sports & Exercise* 32 (2000): 432–8.

Mueller, E. R., L. Borrud, P. S. Goode, S. Meikle, E. R. Mueller, A. Tuteja, A. Weidner, M. Weinstein, and W. Ye. "Fecal Incontinence in US Adults: Epidemiology and Risk Factors." *Gastroenterology* 137 (2009): 512–7.

Murphy, G. S., L. Bodhidatta, P. Echeverria, S. Tansuphaswadikul, C. W. Hoge, S. Imlarp, and K. Tamura. "Ciprofloxacin and Loperamide in the Treatment of Bacillary Dysentery." *Annals of Internal Medicine* 118 (1993): 582–6.

Nelson, P. A. "Probiotics for Treatment of *Clostridium difficile*-associated Colitis in Adults [review]." The Cochrane Library, 2009. Issue 3: 1–16.

Nelson, R. L. "Epidemiology of Fecal Incontinence." *Gastroenterology* 126 (supplement 1, 2004): S3–7.

Nygaard, I., M. D. Barber, K. L. Burgio, K. Kenton, S. Meikle, J. Schaffer, C. Spino, W. E. Whitehead, J. Wu, and D. J. Brody. "Prevalence of Symptomatic Pelvic Floor Disorders in US Women." *Journal of the American Medical Association* 300 (2008): 1311–6.

Petruccelli, B. P., G. S. Murphy, J. L. Sanchez, S. Walz, R. DeFraites, J. Gelnett, R. L. Haberberger, P. Echeverria, and D. N. Taylor. "Treatment of Traveler's Diarrhea with Ciprofloxacin and Loperamide." *Journal of Infectious Diseases* 165 (1992): 557–60.

Pillai, A., and R. Nelson. "Probiotics for Treatment of *Clostridium difficile*-associated Colitis in Adults." *Cochrane Database of Systemic Reviews,* 1 (2008).

Reddy, S. I., and J. L. Wolf. "Acute Diarrhea." In *Office Practice of Medicine.* 4th ed., edited by W. T. Branch, 343–5. Philadelphia; WB Saunders, 2003.

Rubio-Tapia, A., R. A. Kyle, E. L. Kaplan, D. R. Johnson, W. Page, F. Erdtmann, T. L. Brantner, et al. "Increased Prevalence and Mortality in Undiagnosed Celiac Disease." *Gastroenterology* 137 (2009): 88–93.

Rubio-Tapia, A., and J. A. Murray. "Celiac Disease Beyond the Gut [editorial]." *Clinical Gastroenterology and Hepatology* 6 (2008): 722–23.

Sartour, R. B. "Probiotics for Gastrointestinal Disease." www.uptodate.com. Topic updated May 18, 2009.

Sazawal, S., G. Hiremath, U. Dhingra, P. Malik, S. Deb, and R. E. Black. "Efficacy of Probiotics in Prevention of Acute Diarrhoea: A Meta-analysis of Masked, Randomised, Placebo-controlled Trials." *Lancet Infectious Diseases* 6 (2006): 374–82.

Scharff, R. L. "Health-Related Costs from Foodborne Illness in the United States." www.producesafetyproject.org (2010).

Schuppan, D., and W. Dieterich. "Pathogenesis, Epidemiology, and Clinical Manifestations of Celiac Disease." www.uptodate.com. Topic updated October 9, 2009.

Schuppan, D., and Y. Junker. "Turning Swords into Plowshares: Transglutaminase to Detoxify Gluten [editorial]". *Gastroenterology* 133 (2007): 1025–38.

Sciarretta, G., A. Furno, M. Mazzoni, and P. Malaguti. "Post-cholecystectomy Diarrhea: Evidence of Bile Acid Malabsorption Assessed by SeHCAT Test." *American Journal of Gastroenterology* 87 (1992) 1852–4.

Silano, M., C. Agostoni, and S. Guandalini. "Effect of the Timing of Gluten Introduction on the Development of Celiac Disease." *World Journal of Gastroenterology* 16 (April 2010): 1939–42.

Taylor, D. N., J. L. Sanchez, W. Candler, S. Thornton, C. McQueen, and P. Echeverria. "Treatment of Travelers' Diarrhea: Ciprofloxacin Plus Loperamide Compared with Ciprofloxacin Alone. A Placebo-controlled, Randomized Trial." *Annals of Internal Medicine* 114 (1991): 731–4.

Verdu, E. F., D. Armstrong, and J. A. Murray. "Between Celiac Disease and Irritable Bowel Syndrome: The 'No Man's Land' of Gluten Sensitivity." *American Journal of Gastroenterology* 104 (2009): 1587–94.

Vrabie, R., and F. N. Aberra. "Prescribing an Antibiotic? Do Not Forget the Probiotic. [selected summaries]." *Gastroenterology* 137 (2009): 1846–7.

http://www.mayoclinic.com/health/gluten-free-diet/DG00063

http://www.gicare.com/diets/Gluten-Free.aspx

Chapter 5: When You Just Can't Go: Constipation

"American Gastroenterological Association Medical Position Statement: Diagnosis and Treatment of Hemorrhoids." *Gastroenterology* 126 (2004): 1461–2.

"American Gastroenterological Association Technical Review on the Diagnosis and Treatment of Hemorrhoids." *Gastroenterology* 126 (2004): 1463–73.

Anderson, J. W., P. Baird, R. H. Davis Jr., M. Knudtson, A. Koraym, V. Waters, and C. L. Williams. "Health Benefits of Dietary Fiber." *Nutrition Reviews* 67 (2009): 188–205.

Bassotti, G., F. Ghistolini, F. Sietchiping-Nzepa, G. de Roberto, A. Morelli, and G. Chiarioni. "Biofeedback for Pelvic Floor Dysfunction in Constipation." *British Medical Journal* 328 (2004): 393–6.

Bharucha, A. E., A. Wald, P. Enck, and S. Rao. "Functional Anorectal Disorders." *Gastroenterology* 130 (2006): 1510–8.

Bijkerk, C. J., N. J. de Wit, J. W. M. Muris, P. J. Whorwell, J. A. Knotterus, and A. W. Hoes. "Soluble or Insoluble Fibre in Irritable Bowel Syndrome in Primary Care? Randomised Placebo Controlled Trial." *British Medical Journal* 339 (2009): b3154.

Brown, S. R., P. A. Cann, and N. W. Read. "Effect of Coffee on Distal Colon Function." *Gut* 31 (1990): 450–3.

Drossman, D. A., C. B. Morris, Y. Hu, B. B. Toner, N. Diamant, J. Leserman, M. Shetzline, C. Dalton, and S. I. Bangdiwala. "A Prospective Assessment of Bowel Habit in Irritable Bowel Syndrome in Women: Defining an Alternator." *Gastroenterology* 128 (2005): 580–9.

Goel, V., B. Ooraikul and T. K. Basu. "Cholesterol Lowering Effects of Rhubarb Stalk Fiber in Hypercholesterolemic Men." *Journal of the American College of Nutrition* 16 (6) (1997): 600–4.

Kamm, M. A., M. J. Farthing, J. E. Lennard-Jones, L. A. Perry, and T. Chard. "Steroid Hormone Abnormalities in Women with Severe Idiopathic Constipation." *Gut* 32 (1991): 80–4.

Longstreth, G. F., W. G. Thompson, W. D. Chey, L. A. Houghton, F. Mearin, and R. C. Spiller. "Functional Bowel Disorders." *Gastroenterology* 130 (2006): 1480–91.

McCallum, I. J. D., S. Ong, and M. Mercer-Jones. "Chronic Constipation in Adults." *British Medical Journal* 338 (2009): b831.

Müller-Lissner, S. A., M. A. Kamm, C. Scarpignato, and A. Wald. "Myths and Misconceptions about Chronic Constipation." *American Journal of Gastroenterology* 100 (2005): 232–42.

Palsson, O. S., S. Heymen, and W. E. Whitehead. "Biofeedback Treatment for Functional Anorectal Disorders: A Comprehensive Efficacy Review." *Applied Psychophysiology and Biofeedback* 29 (2004): 153–74.

Pizzetti, D., R. Annibali, A. Bufo, and M. Pescatori. "Colonic Hydrotherapy for Obstructed Defecation. [correspondence]." *Colorectal Disease* 7 (2005): 107–8.

Rao, S. S. C., K. Seaton, M. Miller, K. Brown, I. Nygaard, P. Stumbo, B. Zimmerman, and K. Schulze. "Randomized Controlled Trial of Biofeedback, Sham Feedback, and Standard Therapy for Dyssynergic Defecation." *Clinical and Gastroenterology Hepatology* 5 (2007): 331–8.

Richards, D. G., D. L. McMillan, E. A. Mein, and C. D. Nelson. "Colonic Irrigations: A Review of the Historical Controversy and the Potential for Adverse Effects." *Journal of Alternative and Complementary Medicine* 12 (2006): 389–93.

Sagami, Y., Y. Shimada, J. Tayama, T. Nomura, M. Satake, Y. Endo, et al. "Effect of a Corticotropin Releasing Hormone Receptor Antagonist on Colonic Sensory and Motor Function in Patients with Irritable Bowel Syndrome." *Gut* 53 (2004): 958–64.

Sanjeevi, A., and D. F. Kirby. "The Role of Food and Dietary Intervention in the Irritable Bowel Syndrome." *Practical Gastroenterology* 32 (2008): 33–42.

Schiller, L. R. "Nutrients and Constipation: Cause or Cure?" *Practical Gastroenterology* 32 (2008): 43–9.

Taffinder, N. J., E. Tan, I. G. Webb, and P. J. McDonald. "Retrograde Commercial Colonic Hydrotherapy." *Colorectal Disease* 6 (2004): 258–60.

Talley, N. J. "How to Do and Interpret a Rectal Examination in Gastroenterology." *American Journal of Gastroenterology* 103 (2008): 820–2.

http://www.feinberg.northwestern.edu/nutrition/factsheets/fiber.htm

http://www.huhs.harvard.edu/assets/File/OurServices/Service_Nutrition_Fiber.pdf

http://www.hsph.harvard.edu/nutritionsource/what-should-you-eat/fiber/

http://www.tufts.edu/med/nutrition-infection/hiv/health_fiber.html

http://www.medicinenet.com/laxatives_for_constipation

http://www.aapb.org

http://www.wehealthy.org/healthinfo/dietaryfiber/fibercontentechart.html

Chapter 6: Stinky Burps: Heartburn and Halitosis

"AGA Technical Review on the Diagnosis and Treatment of Gastroparesis." *Gastroenterology* 127 (2004): 1592–622.

Allescher, H. D. "Globus Sensation and Hyperdynamic Upper Esophageal Sphincter: Another Piece in the Puzzle? [selected summaries]." *Gastroenterology* 137 (2009): 1847–9.

"American Gastroenterological Association Institute Technical Review on the Management of Gastroesophageal Reflux Disease." *Gastroenterology* 135 (2008): 1392–413.

Asano, K., and H. Suzuki. "Silent Acid Reflux and Asthma Control." *New England Journal of Medicine* 360 (2009): 1551–3.

Asthma Clinical Research Centers. "Efficacy of Esomeprazole for Treatment of Poorly Controlled Asthma." *New England Journal of Medicine* 360 (2009): 1487–99.

Bardhan, K. D., S. Muller-Lissner, M. A. Bigard, G. B. Porro, J. Ponce, J. Hosie, M. Scott, et al. "Symptomatic Gastro-oesophageal Reflux Disease:

Double Blind Controlled Study of Intermittent Treatment with Omeprazole or Ranitidine." *British Medical Journal* 318 (1999): 502–7.

Calil, C. M., P. O. Lima, C. F. Bernardes, F. C. Groppo, F. Bado, and F. K. Marcondes. "Influence of Gender and Menstrual Cycle on Volatile Sulphur Compounds Production." *Archives of Oral Biology* 53 (2008): 1107–12.

Camilleri, M. "Pathogenesis of Delayed Gastric Emptying." www.uptodate.com. Topic updated June 8, 2007.

Camilleri, M. "Treatment of Delayed Gastric Emptying." www.uptodate.com. Topic updated July 11, 2008.

Camilleri, M., D. Dubois, B. Coulie, M. Jones, P. J. Kahrilas, A. M. Rentz, A. Sonnenberg, et al. "Prevalence and Socioeconomic Impact of Upper Gastrointestinal Disorders in the United States: Results of the US Upper Gastrointestinal Study." *Clinical Gastroenterology and Hepatology* 3 (2005): 543–52.

Chey, W. D., and B. C. Y. Wong and the Practice Parameters Committee of the American College of Gastroenterology. "American College of Gastroenterology Guideline on the Management of *Helicobacter pylori* Infection." *American Journal of Gastroenterology* 102 (2007): 1808–25.

Ciovica, R., M. Gadenstatter, A. Klingler, W. Lechner, O. Riedl, and G. P. Schwab. "Quality of Life in GERD Patients: Medical Treatment Versus Antireflux Surgery." *Journal of Gastrointestinal Surgery* 10 (2006): 934–9.

Corley, D. A., A. Kubo, and W. Zhao. "Abdominal Obesity, Ethnicity and Gastro-oesophageal Reflux Symptoms." *Gut* 56 (2007): 756–62.

DeVault, K. R., and D. O. Castell. "Updated Guidelines for the Diagnosis and Treatment of Gastroesophageal Reflux Disease." *American Journal of Gastroenterology* 100 (2005): 190–200.

Drake, D., and D. Hollander. "Neutralizing Capacity and Cost Effectiveness of Antacids." *Annals of Internal Medicine* 84 (1981): 215–7.

Everhart, J. E., ed. "The Burden of Digestive Diseases in the United States." U.S. Department of Health and Human Services, Public Health Service, National Institutes of Health, National Institute of Diabetes and Digestive and Kidney Diseases. Washington, DC: U.S. Government Printing Office, 2008; NIH Publication No. 09-6443 [p. 70].

Falk, G. W. "Obesity and Gastroesophageal Reflux Disease: Another Piece of the Puzzle. [selected summaries]." *Gastroenterology* 134 (2008): 1620–2.

Fass, R., and D. Sifrin. "Management of Heartburn not Responding to Proton Pump Inhibitors." *Gut* 58 (2009): 295–309.

Freedman, N. D., W. H. Chow, Y. T. Gao, X. O. Shu, B. T. Ji, G. Yang, J. H. Lubin, et al. "Menstrual and Reproductive Factors and Gastric Cancer Risk in a Large Prospective Study of Women." *Gut* 56 (2007): 1671–7.

Gray, S. L., A. Z. LaCroix, J. Larson, J. Robbins, J. A. Cauley, J. E. Manson, and Z. Chen. "Proton Pump Inhibitor Use, Hip Fracture, and Change in Bone Mineral Density in Postmenopausal Women: Results from the Women's Health Initiative." *Archives of Internal Medicine* 170(9) (2010): 765–71.

Hasler, W. L. "Garlic Breath Explained: Why Brushing Your Teeth Won't Help. [selected summaries]." *Gastroenterology* 117 (1999): 1248–9.

Herzig, S. J., M. D. Howell, L. H. Ngo, and E. R. Marcantonio. "Acid-suppressive Medication Use and the Risk for Hospital-acquired Pneumonia." *Journal of the American Medical Association* 301 (2009): 2120–8.

Kahrilas, P. J. "Medical Management of Gastroesophageal Reflux Disease in Adults." www.uptodate.com. Topic updated June 9, 2009.

Kwiatek, M. A., F. Mirza, P. J. Kahrilas, and J. E. Pandolfino. "Hyperdynamic Upper Esophageal Sphincter Pressure: A Manometric Observation in Patients Reporting Globus Sensation." *American Journal of Gastroenterology* 104 (2009): 289–98.

Linsky, A., K. Gupta, E. V. Lawler, J. R. Fonda, and J. A. Hermos. "Proton Pump Inhibitors and Risk for Recurrent *Clostridium difficile* Infection." *Archives of Internal Medicine* 170(9) (2010): 772–8.

Lundell, L., S. Attwood, C. Ell, R. Fiocca, J. P. Galmiche, J. Hatlebakk, T. Lind, and O. Junghard. "Comparing Laparoscopic Antireflux Surgery with Esomeprazole in the Management of Patients with Chronic Gastro-oesophageal Reflux Disease: A 3-year Interim Analysis of the LOTUS Trial." *Gut* 57 (2008): 1207–13.

Lundell, L., P. Miettinen, H. E. Myrvold, S. A. Pedersen, B. Liedman, J. G. Hatlebakk, R. Julkonen, et al. "Continued (5-year) Follow-up of a Randomized Clinical Study Comparing Antireflux Surgery and Omeprazole in Gastroesophageal Reflux Disease." *Journal of the American College of Surgeons* 192 (2001): 172–9.

Mahon, D., M. Rhodes, B. Decadt, A. Hindmarsh, R. Lowndes, I. Beckingham, B. Koo, and R. G. Newcombe. "Randomized Clinical Trial of Laparoscopic Nissen Fundoplication Compared with Proton-Pump Inhibitors for Treatment of Chronic Gastro-oesophageal Reflux." *British Journal of Surgery* 92 (2005): 695–9.

McColl, K. E., and D. Gillen. "Evidence that Proton-Pump Inhibitor Therapy Induces the Symptoms It Is Used to Treat. [editorial]." *Gastroenterology* 137 (2009): 20–2.

Moayyedi, P., and B. C. Delaney. "Measuring Gastroesophageal Reflux Symptoms: Musings from Marrakech [editorial]." *American Journal of Gastroenterology* 100 (2005): 19–20.

Moshkowitz, M., N. Horowitz, M. Leshno, and Z. Halpern. "Halitosis and Gastroesophageal Reflux Disease: A Possible Association." *Oral Diseases* 13 (2007): 581–5.

Nocon, M., J. Labenz, D. Jaspersen, A. Leodolter, K. Richters, M. Vieth, T. Linds, P. Malfertheiner, and S. N. Willich. "Health-related Quality of Life in Patients with Gastro-oesophageal Reflux Disease under Routine Care: 5-year Follow-up Results of the ProGERD Study." *Alimentary Pharmacology & Therapeutics* 29 (2009): 662–8.

Nojkov, B., J. H. Rubenstein, S. A. Adis, M. J. Shaw, R. Saad, J. Rai, B. Weinman, and W. D. Chey. "The Influence of Co-morbid IBS and Psychological Distress on Outcomes and Quality of Life following PPI Therapy in Patients with Gastro-oesophageal Reflux Disease." *Alimentary Pharmacology & Therapeutics* 27 (2008): 473–82.

O'Donoghue, M. L., E. Braunwald, E. M. Antman, S. A. Murphy, E. R. Bates, Y. Rozenman, A. D. Michelson, et al. "Pharmacodynamic Effect and Clinical Efficacy of Clopidogrel and Prasugrel with or without a Proton-pump Inhibitor: An Analysis of Two Randomized Trials." *The Lancet* 374 (2009): 989–97.

Reimer, C., B. Søndergaard, L. Hilsted, and P. Bytzer. "Proton-pump Inhibitor Therapy Induces Acid-related Symptoms in Healthy Volunteers after Withdrawal of Therapy." *Gastroenterology* 137 (2009): 80–7.

Robinson, M., F. Lanza, D. Avner, and M. Haber. "Effective Maintenance Treatment of Reflux Esophagitis with Low-dose Lansoprazole: A Randomized, Double-blind, Placebo-controlled Trial." *Annals of Internal Medicine* 124 (1996): 859–67.

Rude, M. K., and W. D. Chey. "Proton-pump Inhibitors, Clopidogrel and Cardiovascular Adverse Events: Fact, Fiction, or Something in Between? [selected summaries]." *Gastroenterology* 137 (2009): 1168–71.

Sachs, G., J. M. Shin, and C. W. Howden. "Review Article: The Clinical Pharmacology of Proton Pump Inhibitors." *Alimentary Pharmacology & Therapeutics* 23 (supplement 2, 2006): 2–8.

Simon, T., C. Verstuyft, M. Mary-Krause, L. Quteineh, E. Drouet, N. Méneveau, P. Gabriel Steg, J. Ferrières, N. Danchin, and L. Becquemont. "Genetic Determinants of Response to Clopidogrel and Cardiovascular Events." *New England Journal of Medicine* 360 (2009): 363–75.

Souza, R. F., and S. J. Spechler. "Concepts in the Prevention of Adenocarcinoma of the Distal Esophagus and Proximal Stomach." *CA: A Cancer Journal for Clinicians* 55 (2005): 334–51.

Spechler, S. J., E. Lee, D. Ahnen, R. K. Goyal, I. Hirano, F. Ramirez, J. P. Raufman, et al. "Long-term Outcome of Medical and Surgical Therapies for Gastroesophageal Reflux Disease: Follow-up of a Randomized Controlled Trial." *Journal of the American Medical Association* 285 (2001): 2331–8.

Suarez, F., J. Springfield, J. Furne, and M. Levitt. "Differentiation of Mouth Versus Gut as Site of Origin of Odoriferous Breath Gases after Garlic Ingestion." *American Journal of Physiology* 276 (1999): G425–30.

Talley, N. J., G. R. Locke, M. McNally, C. D. Schleck, A. R. Zinsmeister, and L. J. Melton III. "Impact of Gastroesophageal Reflux on Survival in the Community." *American Journal of Gastroenterology* 103 (2008): 12–9.

Van den Broek, A. M., L. Feenstra, and C. de Baat. "A Review of the Current Literature on Aetiology and Measurement Methods of Halitosis." *Journal of Dentistry* 35 (2007): 627–35.

Wang, C., Y. Yuan, and R. H. Hunt. "The Association between *Helicobacter pylori* Infection and Early Gastric Cancer: A Meta-analysis." *American Journal of Gastroenterology* 102 (2007): 1789–98.

Wolfe, M. M. "Overview and Comparison of the Proton Pump Inhibitors for the Treatment of Acid-related Disorders." www.uptodate.com. Topic updated October 7, 2009.

Yang, L., X. Lu, C. W. Nossa, F. Francois, R. M. Peek, and Z. Pei. "Inflammation and Intestinal Metaplasia of the Distal Esophagus Are Associated with Alterations in the Microbiome." *Gastroenterology* 137 (2009): 588–97.

Yang, Y. X., J. D. Lewis, S. Epstein, and D. C. Metz. "Long-term Proton Pump Inhibitor Therapy and Risk of Hip Fracture." *Journal of the American Medical Association* 296 (2006): 2947–53.

Chapter 7: When It's Really Bad: Time to Get Help

ACS: Statistics for 2009. Cancer_Statistic_2009_Slides_rev.ppt. http://www.cancer.org.

Afdhal, N. H. "Epidemiology of and Risk Factors for Gallstones." www.uptodate.com. Topic updated October 6, 2009.

Attili, A. F., A. De Santis, R. Capri, A. M. Repice, S. Maselli, and the GREPCO Group. "The Natural History of Gallstones: The GREPCO Experience." *Hepatology* 21 (1995): 655–60.

Barbara, L., C. Sama, A. M. Morselli Labate, F. Taroni, A. G. Rusticali, D. Festi, C. Sapio, et al. "A Population Study on the Prevalence of Gallstone Disease: The Sirmione Study." *Hepatology* 7 (1987): 913–7.

Carter, M. J., A. J. Lobo, and S. P. Travis. "Guidelines for the Management of Inflammatory Bowel Disease in Adults." *Gut* 53 (supplement 5, 2004): V1–16.

Davies, R. J., B. I. O'Connor, C. Victor, H. M. MacRae, Z. Cohen, and R. S. McLeod. "A Prospective Evaluation of Sexual Function and Quality of Life after Ileal Pouch-anal Anastomosis." *Diseases of the Colon & Rectum* 51 (2008): 1032–5.

De Rooy, E. C., B. B. Toner, R. G. Maunder, G. R. Greenberg, D. Baron, A. H. Steinhart, R. McLeod, and Z. Cohen. "Concerns of Patients with Inflammatory Bowel Disease: Results from a Clinical Population." *American Journal of Gastroenterology* 96 (2001): 1816–21.

Dominitz, J. A., and D. J. Robertson. "Colorectal Cancer Screening with Computed Tomographic Colonography [selected summaries]." *Gastroenterology* 136 (2009): 1451–3.

Everhart, J. E., ed. "The Burden of Digestive Diseases in the United States." U.S. Department of Health and Human Services, Public Health Service, National Institutes of Health, National Institute of Diabetes and Digestive and Kidney Diseases. Washington, DC: U.S. Government Printing Office, 2008; NIH Publication No. 09-6443 [pp. 115–6].

Everhart, J. E., M. Khare, M. Hill, and K. R. Maurer. "Prevalence and Ethnic Differences in Gallbladder Disease in the United States." *Gastroenterology* 117 (1999): 632–9.

Hanauer, S. B. "Positioning Biologic Agents in the Treatment of Crohn's Disease." *Inflammatory Bowel Diseases* 15 (2009): 1570–82.

Isaacs, K., and H. Herfarth. "Role of Probiotic Therapy in IBD." *Inflammatory Bowel Diseases* 14 (2008): 1597–605.

Johnson, C. D., M. H. Chen, A. Y. Toledano, J. P. Heiken, A. Dachman, M. D. Kuo, C. O. Menias, et al. "Accuracy of CT Colonography for Detection of Large Adenomas and Cancer." *New England Journal of Medicine* 359 (2008): 1207–17.

Kane, S. "Gender Issues in Inflammatory Bowel Disease." *Women's Health* 1 (2005): 401–7.

Kornbluth, A., and D. B. Sachar. "Ulcerative Colitis Practice Guidelines in Adults (Update): American College of Gastroenterology, Practice Parameters Committee." *American Journal of Gastroenterology* 99 (2004): 1371–85.

Larson, D. W., M. M. Davies, E. J. Dozois, R. R. Cima, K. Piotrowicz, K. Anderson, S. A. Barnes, et al. "Sexual Function, Body Image, and Quality of Life after Laparoscopic and Open Ileal Pouch-Anal Anastomosis." *Diseases of the Colon & Rectum* 51 (2008): 392–6.

Levin, B., D. A. Lieberman, B. McFarland, K. S. Andrews, D. Brooks, J. Bond, C. Dash, et al. "Screening and Surveillance for the Early Detection of Colorectal Cancer and Adenomatous Polyps, 2008: A Joint Guideline from the American Cancer Society, the US Multi-Society Task Force on Colorectal Cancer, and the American College of Radiology." *Gastroenterology* 130 (2006): 940–87.

Lichtenstein, G. R., M. T. Abreu, R. Cohen, and W. Tremaine. "American Gastroenterological Association Institute Medical Position Statement on Corticosteroids, Immunomodulators, and Infliximab in Inflammatory Bowel Disease." *Gastroenterology* 130 (2006): 935–39.

Lichtenstein, G. R., M. T. Abreu, R. Cohen, and W. Tremaine. "American Gastroenterological Association Institute Technical Review on Corticosteroids, Immunomodulators, and Infliximab in Inflammatory Bowel Disease." *Gastroenterology* 130 (2006): 940–87.

Lichtenstein, G. R., S. B. Hanauer, and W. J. Sandborn. "Management of Crohn's Disease in Adults." *American Journal of Gastroenterology* 104 (2009): 465–83.

Lieberman, D. A., J. L. Holub, M. D. Moravee, G. M. Eisen, D. Peters, and C. D. Morris. "Prevalence of Colon Polyps Detected by Colonoscopy Screening in Asymptomatic Black and White Patients." *Journal of the American Medical Association* 300 (2008): 1417–22.

Loftus, E. V. Jr. "Clinical Epidemiology of Inflammatory Bowel Disease: Incidence, Prevalence, and Environmental Influences." *Gastroenterology* 126 (2004): 1504–17.

Louis, E., V. Michel, J. P. Hugot, C. Reenaers, F. Fontaine, M. Delforge, F. El Yafi, J. F. Colombel, and J. Belaiche. "Early Development of Stricturing or Penetrating Pattern in Crohn's Disease Is Influenced by Disease Location, Number of Flares, and Smoking but not by NOD2/CARD15 Genotype." *Gut* 52 (2003): 552–7.

Moody, G., and J. Mayberry. "Perceived Sexual Dysfunction amongst Patients with Inflammatory Bowel Disease." *Digestion* 54 (1993): 256–60.

Moody, G., C. S. Probert, E. M. Srivastava, J. Rhodes, and J. F. Mayberry. "Perceived Sexual Dysfunction amongst Women with Crohn's Disease: A Hidden Problem." *Digestion* 52 (1992): 179–83.

Reddy, S. I., and J. L. Wolf. "Gender Considerations in Inflammatory Bowel Disease." In *Principles of Gender-Specific Medicine*, edited by M. Legato. 428–37. San Diego: Elsevier Academic Press, 2004.

Reddy, S. I., and J. L. Wolf. "Management Issues in Women with Inflammatory Bowel Disease." *Journal of the American Osteopathic Association* 101 (2001): S17–S23.

Regueiro, M. D. "Diagnosis and Treatment of Ulcerative Proctitis." *Journal of Clinical Gastroenterology* 38 (2004): 733–40.

Sheil, B., F. Shanahan, and L. O'Mahony. "Probiotic Effects on Inflammatory Bowel Disease." *Journal of Nutrition* 137 (2007): S819-24.

Travis, S. P., E. F. Stange, M. Lemann, T. Öresland, Y. Chowers, A. Forbes, G. D'Haens, et al. "European Evidence Based Consensus on the Diagnosis and Management of Crohn's Disease: Current Management." *Gut* 55 (supplement 1, 2006): i16–35.

U.S. Preventive Services Task Force. "Screening for Colorectal Cancer: U.S. Preventive Services Task Force Recommendation Statement." *Annals of Internal Medicine* 149 (2008): 627–37.

Van Limbergen, J., D. C. Wilson, and J. Satsangi. "The Genetics of Crohn's Disease." *Annual Review of Genomics and Human Genetics* 10 (2009): 89–116.

Wolf, J. L. "Colorectal Cancer." In *The Savvy Woman Patient: How and Why Sex Differences Affect Your Health*, edited by P. Greenberg and J. Wider, 122–6. Herndon: VA Capital Books Inc., 2006.

Wolf, J. L. "Uniquely Women's Issues in Colorectal Cancer Screening." *American Journal of Gastroenterology* 101 (2006): S625–9.

Wolf, J. L. "Unmasking the Impact of Gender on Inflammatory Bowel Disease." *Women's Health* 1 (2005): 299–303.

Chapter 8: Nine Months of This? Minimizing Stomach Problems During Pregnancy

Acs, N., F. Banhidy, E. H. Puho, and A. E. Czeizel. "Senna Treatment in Pregnant Women and Congenital Abnormalities in Their Offspring—A Population-based Case-control Study." In *Drugs in Pregnancy and Lactation*, 8th ed, edited by G. G. Briggs, R. K. Freeman, and S. J. Yaffe. Philadelphia: Lippincott Williams and Wilkins, 2008.

Chen, M. M., F. V. Coakley, A. Kaimal, and R. K. Laros Jr. "Guidelines for Computed Tomography and Magnetic Resonance Imaging Use During Pregnancy and Lactation." *Obstetrics & Gynecology* 112 (2008): 333–40.

Dubinsky, M., B. Abraham, and U. Mahadevan. "Management of the Pregnant IBD Patient." *Inflammatory Bowel Diseases* 14 (2008): 1736–50.

FDA News. "FDA Requires Boxed Warning and Risk Mitigation Strategy for Metoclopramide-containing Drugs." February 26, 2009.

Gill, S. K., L. O'Brien, T. R. Einarson, and G. Koren. "The Safety of Proton Pump Inhibitors (PPIs) in Pregnancy: A Meta-analysis." *American Journal of Gastroenterology* 104 (2009): 1541–5.

Keller, J., D. Frederking, and P. Layer. "The Spectrum and Treatment of Gastrointestinal Disorders during Pregnancy." *Nature Clinical Practice Gastroenterology & Hepatology* 5 (2008): 430–43.

Knight, B., C. Mudge, S. Openshaw, A. White, and A. Hart. "Effect of Acupuncture on Nausea of Pregnancy: A Randomized, Controlled Trial." *Obstetrics & Gynecology* 97 (2001): 184–8.

Lawson, M., F. Kern Jr., and G. T. Everson. "Gastrointestinal Transit Time in Human Pregnancy: Prolongation in the Second and Third Trimesters Followed by Postpartum Normalization." *Gastroenterology* 89 (1985): 996–9.

Levy, N., E. Lemberg, and M. Sharf. "Bowel Habit in Pregnancy." *Digestion* 4 (1971): 216–22.

Ludvigsson, J. F., S. M. Montgomery, and A. Ekbom. "Celiac Disease and Risk of Adverse Fetal Outcome: A Population-based Cohort Study." *Gastroenterology* 129 (2005): 454–63.

Mahadevan, U., and S. Kane. "American Gastroenterological Association Institute Technical Review on the Use of Gastrointestinal Medications in Pregnancy." *Gastroenterology* 131 (2006): 283–311.

Matok, I., R. Gorodischer, G. Koren, E. Sheiner, A. Wiznitzer, and A. Levy. "The Safety of Metoclopramide Use in the First Trimester of Pregnancy." *New England Journal of Medicine* 360 (2009): 2528–35.

Misri, S., and S. I. Lusskin. "Management of Depression in Pregnant Women." www.update.com. Topic updated June 18, 2010.

Olans, L. B., and J. L. Wolf. "Gastroesophageal Reflux in Pregnancy." *Gastrointestinal Endoscopy Clinics of North America* 4 (1994): 699–712.

Ozgoll, G., M. Goll, and M. Simbar. "Effects of Ginger Capsules on Pregnancy, Nausea, and Vomiting." *Journal of Alternative and Complementary Medicine* 15 (2009): 243–6.

Qureshi, W. A., E. Rajan, D. G. Adler, R. E. Davila, W. K. Hirota, B. C. Jacobson, J. A. Leighton, et al. "ASGE Guideline: Guidelines for Endoscopy in Pregnant and Lactating Women." *Gastrointestinal Endoscopy* 61 (2005): 357–62.

Rey, E., F. Rodriguez-Artalejo, A. Herraiz, P. Sanchez, A. Alvarez-Sanchez, M. Escudero, and M. Diaz-Rubio. "Gastroesophageal Reflux Symptoms during and after Pregnancy: A Longitudinal Study." *American Journal of Gastroenterology* 102 (2007): 2395–400.

Rosen, T., M. de Veciana, H. S. Miller, L. Stewart, A. Rebarber, and R. N. Slotnick. "A Randomized Controlled Trial of Nerve Stimulation for Relief of Nausea and Vomiting in Pregnancy." *Obstetrics & Gynecology* 102 (2003): 129–35.

Streitberger, K., J. Ezzo, and A. Schneider. "Acupuncture for Nausea and Vomiting: An Update of Clinical and Experimental Studies." *Autonomic Neuroscience* 129 (2006): 107–17.

Tata, L. J., T. R. Card, R. F. Logan, R. B. Hubbard, C. J. Smith, and J. West. "Fertility and Pregnancy-related Events in Women with Celiac Disease: A Population-based Cohort Study." *Gastroenterology* 128 (2005): 849–55.

Thukral, C., and J. L. Wolf. "Therapy Insight: Drugs for Gastrointestinal Disorders in Pregnant Women." *Nature Clinical Practice Gastroenterology & Hepatology* 3 (2006): 256–66.

Wald, A., D. H. Van Thiel, L. Hoechstetter, J. S. Gavaler, K. M. Egler, R. Verm, L. Scott, and R. Lester. "Effect of Pregnancy on Gastrointestinal Transit." *Digestive Diseases and Sciences* 27 (1982): 1015–8.

Chapter 9: Eating Your Way to Health

Anderson, J. W., P. Baird, R. H. Davis Jr., S. Ferreri, M. Knudtson, A. Koraym, V. Waters, and C. L. Williams. "Health Benefits of Dietary Fiber." *Nutrition Reviews* 67 (2009): 188–205.

Boden, G. "High- or Low-carbohydrate Diets: Which Is Better for Weight Loss, Insulin Resistance, and Fatty Livers?" *Gastroenterology* 136 (2009): 1490–2.

Campbell, S. M. "Hydration Needs Throughout the Lifespan." *Journal of the American College of Nutrition* 26 (2007): S585-87.

Cheuvront, S. N. "The Zone Diet Phenomenon: A Closer Look at the Science behind the Claims." *Journal of the American College of Nutrition* 22 (2003): 9–17.

Dansinger, M. L., J. A. Gleason, J. L. Griffith, H. P. Selker, and E. J. Schaefer. "Comparison of the Atkins, Ornish, Weight Watchers, and Zone Diets for Weight Loss and Heart Disease Risk Reduction: A Randomized Trial." *Journal of the American Medical Association* 293 (2005): 43–53.

Davy, B. M., E. A. Dennis, A. L. Dengo, K. L. Wilson, and K. P. Davy. "Water Consumption Reduces Energy Intake at a Breakfast Meal in Obese Older Adults." *Journal of the American Dietetic Association* 108 (2008): 1236–9.

Drisko, J., B. Bischoff, M. Hall, and R. McCallum. "Treating Irritable Bowel Syndrome with a Food Elimination Diet followed by Food Challenge and Probiotics." *Journal of the American College of Nutrition* 25 (2006): 514–22.

Eckel, R. H. "The Dietary Approach to Obesity: Is it the Diet or the Disorder?" *Journal of the American Medical Association* 293 (2005): 96–7.

Elango, R., R. O. Ball, and P. B. Penchrz. "Individual Amino Acid Requirements in Humans: An Update." *Current Opinion in Clinical Nutrition & Metabolic Care* 11 (2008): 34–9.

Ervin, R. B. "Healthy Eating Index Scores among Adults, 60 Years of Age and Over, by Sociodemographic and Health Characteristics: United States, 1999–2002." Advance Data from Vital and Health Statistics, no. 395. Hyattsville, MD: National Center for Health Statistics, 2008.

Eyre, H., R. Kahn, R. M. Robertson, and the ACS/ADA/AHA Collaborative Writing Committee. "Preventing Cancer, Cardiovascular Disease, and Diabetes: A Common Agenda for the American Cancer Society, the American Diabetes Association, and the American Heart Association." *Circulation* 109 (2004): 3244–55.

Fontana, L., and S. Klein. "Aging, Adiposity, and Calorie Restriction." *Journal of the American Medical Association* 297 (2007): 986–94.

Foreyt, J. P., J. Salas-Salvado, B. Caballero, M. Bulló, K. D. Gifford, I. Bautista, and L. Serra-Majem. "Weight-Reducing Diets: Are There Any Differences?" *Nutrition Reviews* 67 (2009): S99–S101.

Gardner, C. D., A. Kiazand, S. Alhassan, S. Kim, R. S. Stafford, R. R. Balise, H. C. Kraemer, and A. C. King. "Comparison of the Atkins, Zone, Ornish, and LEARN Diets for Change in Weight and Related Risk Factors Among Overweight Premenopausal Women: The A TO Z Weight Loss Study: A Randomized Trial." *Journal of the American Medical Association* 297 (2007): 969–77.

Giovannini, M., E. Verduci, S. Scaglioni, E. Salvatici, M. Bonza, E. Riva, and C. Agostoni. "Breakfast a Good Habit: Not a Repetitive Custom." *Journal of International Medical Research* 36 (2008): 613–24.

Gottschall, E. *The Specific Carbohydrate Diet in Breaking the Vicious Cycle: Intestinal Health Through Diet.* Baltimore, ON: Kirkton Press, Limited, 1994.

Guenther, P. M., W. Y. Juan, J. Reedy, P. Britten, M. Lino, A. Carlson, H. H. Hiza, and S. M. Krebs-Smith. "Diet Quality of Americans in 1994–96 and 2001–02 as Measured by the Healthy Eating Index 2005." *Nutrition Insight* 37 (December 2007).

Heizer, W. D., S. Southern, and S. McGovern. "The Role of Diet in Symptoms of Irritable Bowel Syndrome in Adults: A Narrative Review." *Journal of the American Dietetic Association* 109 (2009): 1204–14.

Hornick, B. A., A. J. Krester, and T. A. Nicklas. "Menu Modeling with MyPyramid Food Patterns: Incremental Dietary Changes Lead to Dramatic Improvements in Diet Quality of Menus." *Journal of the American Dietetic Association* 108 (2008): 2077–83.

"Institute of Medicine (IOM) Dietary Reference Intakes for Energy, Carbohydrate, Fiber, Fat, Fatty Acids, Cholesterol, Protein, and Amino Acids." This report may be accessed via http://www.nap.edu.

Katan, M. B. "Weight-loss Diets for the Prevention and Treatment of Obesity [editorial]." *New England Journal of Medicine* 360 (2009): 923–5.

Kontogianni, M. D., A. Zampelas, and C. Tsigos. "Nutrition and Inflammatory Load." *Annals of the New York Academy of Sciences* 1083 (2006): 214–38.

Lappe, J. M., D. Travers-Gustafson, K. M. Davies, R. R. Recker, and R. P. Heaney. "Vitamin D and Calcium Supplementation Reduces Cancer Risk: Results of a Randomized Trial." *American Journal of Clinical Nutrition* 85 (2007): 1586–91. Erratum in: *American Journal of Clinical Nutrition* 87 (2008): 794.

Lavie, C. J., R. V. Milani, M. R. Mehra, and H. O. Ventura. "Omega-3 Polyunsaturated Fatty Acids and Cardiovascular Diseases." *Journal of the American College of Cardiology* 54 (2009): 585–94.

Rangnekar, A. S., and W. D. Chey. "The FODMAP Diet for Irritable Bowel Syndrome: Food Fad or Roadmap to a New Treatment Paradigm? (Selected Summaries)." *Gastroenterology* 137 (2009): 383–6.

Raybould, H. E. "Nutrient Sensing in the Gastrointestinal Tract: Possible Role for Nutrient Transporters." *Journal of Physiology and Biochemistry* 64 (2008): 349–56.

Reedy, J., and S. M. Krebs-Smith. "A Comparison of Food-based Recommendations and Nutrient Values of Three Food Guides: USDA's MyPyramid, NHLBI's Dietary Approaches to Stop Hypertension Eating Plan and Harvard's Healthy Eating Pyramid." *Journal of the American Dietetic Association* 108 (2008): 522–8.

Sacks, F. M., G. A. Gray, V. J. Carey, S. R. Smith, D. H. Ryan, S. D. Anton, K. McManus, et al. "Comparison of Weight-loss Diets with Different Compositions of Fat, Protein, and Carbohydrates." *New England Journal of Medicine* 360 (2009): 859–73.

Shepherd, S. J., F. C. Parker, J. G. Muir, and P. R. Gibson. "Dietary Triggers of Abdominal Symptoms in Patients with Irritable Bowel Syndrome: Randomized Placebo-controlled Evidence." *Clinical Gastroenterology and Hepatology* 6 (2008): 765–71.

Slavin, J. L., V. Savarino, A. Paredes-Diaz, and G. Fotopoulos. "A Review of the Role of Soluble Fiber in Health with Specific Reference to Wheat Dextrin." *Journal of International Medical Research* 37 (2009): 1–17.

Soenen, S., and M. S. Westerterp-Plantenga. "Proteins and Satiety: Implications for Weight Management." *Current Opinion in Clinical Nutrition & Metabolic Care* 11 (2008): 747–51.

Timlin, M. T., M. A. Pereira, M. Story, and D. Neumark-Sztainer. "Breakfast Eating and Weight Change in a 5-year Prospective Analysis of Adolescents: Project EAT (Eating among Teens)." *Pediatrics* 121 (2008): e638–45.

U.S. Department of Health and Human Services and U.S. Department of Agriculture. *Dietary Guidelines for Americans,* 2005. 6th ed. Washington, DC: U.S. Government Printing Office, 2005.

This publication, as well as the booklet *Finding Your Way to a Healthier You* may be viewed and downloaded from the Internet at http://www.healthierus. gov/dietaryguidelines.

Wactawski-Wende, J., J. M. Kotchen, G. L. Anderson, A. R. Assaf, R. L. Brunner, M. J. O'Sullivan, K. L. Margolis, et al. "Calcium Plus Vitamin D Supplementation and the Risk of Colorectal Cancer." *New England Journal of Medicine* 354 (2006): 684–96.

Westerterp-Plantenga, M. S., A. Nieuwenhuizen, D. Tomé, S. Soenen, and K. R. Westerterp. "Dietary Protein, Weight Loss, and Weight Maintenance." *Annual Review of Nutrition* 29 (2009): 21–41.

Yetley, E. A., D. Brule, M. C. Cheney, C. D. Davis, K. A. Esslinger, P. W. Fischer, K. E. Friedl, et al. "Dietary Reference Intakes for Vitamin D: Justification for a Review of the 1997 Values." *American Journal of Clinical Nutrition* 89 (2009): 719–27.

http://www.hsph.harvard.edu/nutritionsource/what-should-you-eat/pyramid/index.html

http://www.mayoclinic.com/health/gluten-free-diet/DG00063

http://www.celiac.com

http://www.nhlbi.nih.gov/health/public/heart/hbp/dash/new_dash.pdf

http://www.nhlbi.nih.gov/hbp/prevent/h_eating/h_eating.htm

http://www.hsph.harvard.edu/nutritionsource/index.html

http://www.mypyramid.gov/pyramid/index.html

http://www.DietaryGuidelines.gov

http://www.fsis.usda.gov/FactSheets/Focus_On_Freezing/index.asp#12

http://www.healthierus.gov/dietaryguidelines

Weight-loss and Nutrition Myths. How Much Do you Really Know? http://www.niddk.nih.gov.

Fiber Content of Foods: http://www.nal.usda.gov/fnic/foodcomp

Fiber Content of Foods in Common Portions: huhs.harvard.edu/assets/File/OurServices/Service_Nutrition_Fiber.pdf

http://www.cdc.gov/nutrition/everyone/basics/protein.html

Chapter 10: Doctors' Visits and Medications

http://www.fda.gov/Drugs/ResourcesForYou/Consumers/BuyingUsingMedicineSafely/UnderstandingGenericDrugs/ucm167991.htm

http://www.medpagetoday.com/ProductAlert/Prescriptions/15685

http://www.uspharmacist.com/content/s/78/c/13785/

Acknowledgments

I first considered writing a book on intestinal health for women more than ten years ago when I was chairperson of the American Digestive Health Foundation Digestive Health Initiative: Women's Digestive Health. The idea remained at a simmer until my book agent, Gail Ross, convinced me of the burning need for such a project and encouraged me to take it on. She has been a wonderful resource and advocate.

Happily, Gail introduced me to my collaborator, Kara Baskin. Kara has provided the book with perkiness and a contemporary touch. She instinctively knows how to present a dry and even embarrassing topic in a readable, humorous way. Her advice has been invaluable.

The project would have withered on the vine without the enthusiasm and commitment of my editor, Deborah Brody, at Harlequin Enterprises LTD. It has been a joy to work with her.

I will be eternally grateful for the friendship and mentorship of my chief of gastroenterology at Beth Israel Deaconess Medical Center, Dr. J. Thomas Lamont. He has supported me in my clinical and academic pursuits, and he has been my advocate for awards and promotion. He took time out of his exceedingly busy schedule to read the manuscript and offer helpful comments. My gratitude also goes to my colleagues and the staff at Beth Israel Deaconess Medical Center for providing a wonderful working environment and much-needed support when deadlines required relief from my patient schedule. Special thanks to

Dr. Anthony Lembo, a specialist in irritable bowel syndrome, who offered critical advice and corrections; Dr. James Rabb, a gastroenterologist, who offered many suggestions for topics to include; Dr. Martin Smith, a gastrointestinal radiologist, who carefully edited the information on the radiology procedures; and Dr. Mark Zeidel, chief of medicine, who has enthusiastically supported my professional development as a clinician and writer. My administrative assistants, Julie Renz, Siobhan Connolly and Elizabeth Manfredini, provided outstanding ideas and assistance during various aspects of the project.

I would like to thank Dr. Anil Shukla for his professional illustrations. He devoted a considerable portion of his free time to providing these top-notch drawings.

Furthermore, I would like to tip my cap to Emily Lamont for reading and commenting on my manuscript, and to all my friends for their encouragement, helpful suggestions and/or contributions to the book. Thank you to Howard Yoon and others at the Gail Ross Literary Agency who provided helpful critiques of the book proposal.

This book never would have been possible without my remarkable patients, who contributed their personal histories and experiences for this endeavor. It has been my privilege to work with all of them and to be entrusted with their medical care. Unfortunately, only a fraction of the advice and stories they related to me could be included in the book.

Thanks are due to my colleagues and associates at the Society for Women's Health Research for their commitment to discovering the role of sex differences in clinical care and disease prevention for both women and men. This book is a small contribution to that mission.

Finally, I am most indebted to my family. Throughout my career, they have been my biggest advocates and supporters. My father, Louis Wolf, believed that a young woman could enter any field and succeed as well as a man. My mother, Carolyn Sachs, demonstrated that one could successfully open up a business field to and for the benefit of women. My children, Laura and Rebecca, have constantly encouraged my professional pursuits, even when it interfered with family time. They constantly challenged me to answer difficult medical questions in understandable terms. My husband, David Perlman, has always stimulated me to find creative and scientific explanations for medical problems, offered me grounded advice and encouragement, and put up with my long hours in and out of the hospital.

Index

Page numbers of illustrations appear in italics.